T0275982

SpringerBriefs in Statistics

JSS Research Series in Statistics

Editor-in-Chief

Naoto Kunitomo
Akimichi Takemura

Series editors

Genshiro Kitagawa
Tomoyuki Higuchi
Yutaka Kano
Toshimitsu Hamasaki
Shigeyuki Matsui
Manabu Iwasaki
Yasuhiro Omori
Yoshihiro Yajima

The current research of statistics in Japan has expanded in several directions in line with recent trends in academic activities in the area of statistics and statistical sciences over the globe. The core of these research activities in statistics in Japan has been the Japan Statistical Society (JSS). This society, the oldest and largest academic organization for statistics in Japan, was founded in 1931 by a handful of pioneer statisticians and economists and now has a history of about 80 years. Many distinguished scholars have been members, including the influential statistician Hirotugu Akaike, who was a past president of JSS, and the notable mathematician Kiyosi Itô, who was an earlier member of the Institute of Statistical Mathematics (ISM), which has been a closely related organization since the establishment of ISM. The society has two academic journals: the Journal of the Japan Statistical Society (English Series) and the Journal of the Japan Statistical Society (Japanese Series). The membership of JSS consists of researchers, teachers, and professional statisticians in many different fields including mathematics, statistics, engineering, medical sciences, government statistics, economics, business, psychology, education, and many other natural, biological, and social sciences.

The JSS Series of Statistics aims to publish recent results of current research activities in the areas of statistics and statistical sciences in Japan that otherwise would not be available in English; they are complementary to the two JSS academic journals, both English and Japanese. Because the scope of a research paper in academic journals inevitably has become narrowly focused and condensed in recent years, this series is intended to fill the gap between academic research activities and the form of a single academic paper.

The series will be of great interest to a wide audience of researchers, teachers, professional statisticians, and graduate students in many countries who are interested in statistics and statistical sciences, in statistical theory, and in various areas of statistical applications.

More information about this series at http://www.springer.com/series/13497

Yuichi Mori · Masahiro Kuroda
Naomichi Makino

Nonlinear Principal Component Analysis and Its Applications

 Springer

Yuichi Mori
Okayama University of Science
Okayama
Japan

Naomichi Makino
Osaka University
Osaka
Japan

Masahiro Kuroda
Okayama University of Science
Okayama
Japan

ISSN 2191-544X
SpringerBriefs in Statistics
ISSN 2364-0057
JSS Research Series in Statistics
ISBN 978-981-10-0157-4
DOI 10.1007/978-981-10-0159-8

ISSN 2191-5458 (electronic)

ISSN 2364-0065 (electronic)

ISBN 978-981-10-0159-8 (eBook)

Library of Congress Control Number: 2016957863

© The Author(s) 2016
This book was advertised with a copyright holder in the name of the publisher in error, whereas the author holds the copyright.
This work is subject to copyright. All rights are reserved by the Publisher, whether the whole or part of the material is concerned, specifically the rights of translation, reprinting, reuse of illustrations, recitation, broadcasting, reproduction on microfilms or in any other physical way, and transmission or information storage and retrieval, electronic adaptation, computer software, or by similar or dissimilar methodology now known or hereafter developed.
The use of general descriptive names, registered names, trademarks, service marks, etc. in this publication does not imply, even in the absence of a specific statement, that such names are exempt from the relevant protective laws and regulations and therefore free for general use.
The publisher, the authors and the editors are safe to assume that the advice and information in this book are believed to be true and accurate at the date of publication. Neither the publisher nor the authors or the editors give a warranty, express or implied, with respect to the material contained herein or for any errors or omissions that may have been made.

Printed on acid-free paper

This Springer imprint is published by Springer Nature
The registered company is Springer Nature Singapore Pte Ltd.
The registered company address is: 152 Beach Road, #22-06/08 Gateway East, Singapore 189721, Singapore

Preface

As the data size increases and the data structure becomes increasingly complex, considering the nonlinearity, mixed measurement level data, simultaneous analysis and computational efficiency, for example, become increasingly necessary.

Accordingly, we have been investigating data analysis and developing methods related to the above problems: how to jointly analyze data that include not only categorical variables but also numerical variables; matrix operations approaching categorical data analyses; how to select a reasonable subset of variables from mixed measurement level data; analyses based on the sparseness of data and simultaneous estimations, such as a joint method with dimension reduction and clustering, which have recently been investigated; and how to accelerate the computations involving the alternating least squares (ALS) algorithm.

Although we have heretofore been investigating these topics separately, we decided to include the above-mentioned methods/techniques in this series when we started to write this Springer Brief Statistics series.

We present our research interests in two parts as follows:

The first part consists of two chapters that introduce the principles of nonlinear principal component analysis (PCA). After a brief introduction of the ordinary PCA, a PCA for categorical (nominal and ordinal) data is introduced as a nonlinear PCA, in which the optimal scaling technique is used to quantify the categorical variables. The ALS algorithm is the main algorithm used in this method. Next, multiple correspondence analysis (MCA), which is a special case of nonlinear PCA, is introduced, and ALS is also used in the computation. All formulations in these methods are integrated in the same manner as the matrix operations. Since any measurement level data can be treated consistently as numerical data and ALS is a very powerful tool for estimations, these methods can be used in a variety of fields, including biometrics, econometrics, psychometrics, and sociology.

The second part of the book consists of four chapters, which describe applications of the nonlinear PCA: variable selection for mixed measurement level data, sparse MCA, reduced k-means clustering, and an acceleration of the ALS algorithm. The variable selection methods used in PCA, which were originally developed for numerical data, can be applied to any type of measurement level data using nonlinear PCA. Sparseness and k-means clustering for nonlinear data, which were proposed in recent studies, are extensions obtained using the same matrix operations used in nonlinear PCA and MCA. Finally, an acceleration algorithm is proposed to reduce the computational cost of ALS iteration in nonlinear multivariate methods.

This book demonstrates the usefulness of nonlinear PCA and MCA, which can be applied to different measurement level data in a variety of fields and covers the latest topics, including the extension of traditional statistical methods, newly proposed nonlinear methods, and the computational efficiency of these methods. The ALS algorithm is a key concept in all chapters, and optimal quantifications and matrix operations are also key concepts in this book.

We sincerely hope that the concepts and applications presented herein will be of use.

Okayama, Japan   Yuichi Mori
June 2016   Masahiro Kuroda
                                                         Naomichi Makino

Contents

Chapter 1
Introduction

The principles of nonlinear principal component analysis (PCA) and multiple correspondence analysis (MCA), which are useful methods for analyzing mixed measurement level data, and related applications are introduced in this book.

Numerous studies have analyzed data that jointly include categorical (nominal and ordinal) and numerical variables.

Typical methods for the traditional analyses of such data analyze categorical and numerical variables independently, quantify categorical variables by translating them to dummy variables, and use categorical codes as numerical values. These approaches are often not suitable for cases in which we wish to use all variables with a unified scale in order to observe all data to the extent possible, e.g., in lower dimensions.

A good solution is quantification based on optimal scaling. Well-known methods include nonlinear PCA and MCA, which are introduced in the first part of the book. Once we obtain optimal quantification methods for mixed measurement level data, we can apply any of the traditional methods developed for numerical data to mixed measurement level data, for example, to solve the variable selection problem for mixed measurement level data. Moreover, we can propose some extensions to nonlinear PCA and MCA, such as sparse analysis and k-means clustering for nonlinear data, which are based on recent studies and are obtained by the same matrix operations used in nonlinear PCA and MCA. Finally, since effective methods for linearly converging sequences have been developed in the numerical analysis field, we wish to perform alternating least squares (ALS) computations that generate linear convergent sequences more effectively using the results from the field of numerical analysis. These extensions and applications will be illustrated in the second part of the book.

Here, we present an outline of each chapter.

The first part of the book consists of two introductory chapters.

The first chapter (Chap. 2) of Part I introduces nonlinear PCA. PCA is a popular descriptive multivariate method for handling quantitative data and can be extended to deal with mixed measurement level data. For extended PCA with such a mixture

© The Author(s) 2016
Y. Mori et al., *Nonlinear Principal Component Analysis and Its Applications*,
JSS Research Series in Statistics, DOI 10.1007/978-981-10-0159-8_1

of quantitative and qualitative data, we require the quantification of qualitative data to obtain optimal scaling data. PCA with optimal scaling is referred to as nonlinear PCA. Nonlinear PCA, including optimal scaling, alternates between estimating the parameters of PCA and quantifying qualitative data. The ALS algorithm is used for nonlinear PCA and can find least squares solutions by minimizing two types of loss functions: a low-rank approximation and a homogeneity analysis loss function with restrictions. PRINCIPALS and PRINCALS are the ALS algorithms used for the computation of nonlinear PCA.

The second chapter (Chap. 3) of Part I introduces MCA as a special case of the nonlinear PCA. MCA is a widely used technique for analyzing any type of categorical data and aims to reduce large sets of variables into smaller sets of components that summarize the information contained in the data. The purpose of MCA is the same as that of PCA, and MCA can be regarded as an adaptation to categorical data of PCA. There are various approaches to formulating an MCA. We introduce a formulation in which the quantified data matrix is approximated by a lower-rank matrix using the quantification technique.

The second part of the book discusses applications of nonlinear PCA.

The first chapter (Chap. 4) of Part II describes the variable selection in nonlinear PCA. Any measurement level multivariate data can be uniformly treated as numerical data in the context of PCA by the ALS with optimal scaling. This means that all variables of the data can be analyzed as numerical variables, and, therefore, we can solve the variable selection problem for mixed measurement level data using any of the existing variable selection methods developed for numerical variables. In this chapter, we discuss variable selection in nonlinear PCA. We select a subset of variables that represents all variables to the extent possible using the criteria in the modified PCA, which naturally includes the variable selection procedure.

The second chapter (Chap. 5) of Part II illustrates sparse MCA. In MCA, an estimated solution can be transformed into a simple structure in order to simplify the interpretation. The rotation technique is widely used for this purpose. However, an alternative approach, called sparse MCA, has also been proposed. One advantage of sparse MCA is that, in contrast to unrotated or rotated ordinary MCA loadings, some loadings in sparse MCA can be exactly zero. A real data example demonstrates that sparse MCA can provide simple solutions.

The third chapter (Chap. 6) of Part II illustrates the joint dimension reduction and clustering. Although cluster analysis is a technique that aims at dividing objects into similar groups, it does not work properly when variables that do not reflect the clustering structure are included in a dataset or when the number of variables is large. One approach to mitigate this problem is to jointly perform clustering of objects and dimension reduction of variables. In this chapter, we review a technique that combines MCA and k-means clustering to obtain qualitative variables relationships.

In the fourth chapter (Chap. 7) of Part II, an acceleration algorithm of ALS is proposed. As shown in Chap. 2, the nonlinear PCA requires iterative computation using the ALS algorithm (PRINCIPALS/PRINCALS) that alternates between optimal scaling for quantifying qualitative data and analysis of the optimally scaled data using the ordinary PCA approach. When applying nonlinear PCA to very large datasets of numerous nominal and ordinal variables, the ALS algorithm may require many iterations and significant computation time to converge. One reason for the slow convergence of the ALS algorithm is that its speed of convergence is linear. In order to accelerate the convergence of the ALS algorithm, we propose a new iterative algorithm using an idea (the vector ε algorithm) from the field of numerical analysis to generate a faster linear convergent sequence.

This book demonstrates the usefulness of the nonlinear PCA and MCA, which can be applied to different measurement level data in a variety of fields, and considers the latest topics associated with these analyses, including the extension of the traditional statistical method, newly proposed nonlinear methods, and the computational efficiency of these methods. The ALS algorithm is one of the key concepts in all chapters, and optimal quantification and matrix operations are also key concepts in this book.

Part I
Nonlinear Principal Component Analysis

The first part of this book consists of two chapters, which introduce the principle concepts.

In the first chapter (Chap. 2), after a brief introduction of the ordinary principal component analysis (PCA), a PCA for categorical data (nominal and ordinal data) is introduced as nonlinear PCA, in which an optimal scaling technique is used to quantify the categorical variables. The alternating least squares (ALS) algorithm is one of the main algorithms of the proposed method.

In the second chapter (Chap. 3), multiple correspondence analysis, which is a well-known method for categorical variables but is regarded here as a special case of nonlinear PCA, is introduced. The ALS algorithm is also used in the computation. A key point in the first part of this book is that all formulations used in these methods are integrated in the same manner for matrix operations. Since any measurement level data can be treated consistently as numerical data by the methods and ALS is a very powerful tool for estimation, these methods are used in a variety of fields.

Chapter 2
Nonlinear Principal Component Analysis

Abstract Principal components analysis (PCA) is a commonly used descriptive multivariate method for handling quantitative data and can be extended to deal with mixed measurement level data. For the extended PCA with such a mixture of quantitative and qualitative data, we require the quantification of qualitative data in order to obtain optimal scaling data. PCA with optimal scaling is referred to as nonlinear PCA, (Gifi, Nonlinear Multivariate Analysis. Wiley, Chichester, 1990). Nonlinear PCA including optimal scaling alternates between estimating the parameters of PCA and quantifying qualitative data. The alternating least squares (ALS) algorithm is used as the algorithm for nonlinear PCA and can find least squares solutions by minimizing two types of loss functions: a low-rank approximation and homogeneity analysis with restrictions. PRINCIPALS of Young et al. (Principal components of mixed measurement level multivariate data: an alternating least squares method with optimal scaling features 43:279–281, 1978) and PRINCALS of Gifi (Nonlinear Multivariate Analysis. Wiley, Chichester, 1990) are used for the computation.

Keywords Optimal scaling · Quantification · Alternating least squares algorithm · Low-rank approximation · Homogeneity analysis

2.1 Principal Component Analysis

Let $\mathbf{Y} = (\mathbf{y}_1 \ \mathbf{y}_2 \ \ldots \ \mathbf{y}_p)$ be a data matrix of n objects by p numerical variables and each column of $\mathbf{Y}$ be standardized, i.e., $\mathbf{y}_i^\top \mathbf{1}_n = 0$ and $\mathbf{y}_i^\top \mathbf{y}_i / n = 1$ for $i = 1, \ldots, p$, where $\mathbf{1}_n$ is an $n \times 1$ vector of ones.

Principal component analysis (PCA) linearly transforms $\mathbf{Y}$ of p variables into a substantially smaller set of uncorrelated variables that contains much of the information of the original data set. Then PCA simplifies the description of $\mathbf{Y}$ and reveals the structure of $\mathbf{Y}$ and the variables.

PCA postulates that $\mathbf{Y}$ is approximated by the bilinear form

$$\hat{\mathbf{Y}} = \mathbf{Z}\mathbf{A}^\top, \tag{2.1}$$

© The Author(s) 2016
Y. Mori et al., *Nonlinear Principal Component Analysis and Its Applications*,
JSS Research Series in Statistics, DOI 10.1007/978-981-10-0159-8_2

where $\mathbf{Z}$ is an $n \times r$ matrix of n component scores on r $(1 \leq r \leq p)$ components and $\mathbf{A}$ is a $p \times r$ weight matrix that gives the coefficients of the linear combinations. PCA is formulated in terms of the loss function

$$\sigma(\mathbf{Z}, \mathbf{A}) = \text{tr}(\mathbf{Y} - \hat{\mathbf{Y}})^\top (\mathbf{Y} - \hat{\mathbf{Y}}) = \text{tr}(\mathbf{Y} - \mathbf{Z}\mathbf{A}^\top)^\top (\mathbf{Y} - \mathbf{Z}\mathbf{A}^\top). \qquad (2.2)$$

The minimum of the loss function (2.1) over $\mathbf{Z}$ and $\mathbf{A}$ is found by the eigen-decomposition of $\mathbf{Y}^\top \mathbf{Y}/n$ or the singular value decomposition of $\mathbf{Y}$.

2.1.1 Eigen-Decomposition of $\mathbf{Y}^\top \mathbf{Y}/n$

Let $\mathbf{S} = \mathbf{Y}^\top \mathbf{Y}/n$ be a $p \times p$ symmetric matrix. Then we have the following relation between the eigenvalues and eigenvectors of $\mathbf{S}$:

$$\mathbf{S}\mathbf{a}_i = \lambda_i \mathbf{a}_i, \quad \mathbf{a}_i^\top \mathbf{a}_i = 1 \quad \text{and} \quad \mathbf{a}_i^\top \mathbf{a}_j = 0 \ (i \neq j) \qquad (2.3)$$

for $i, j = 1, 2, \ldots, p$. We denote the $p \times p$ matrix having p eigenvectors as columns by $\mathbf{A}$ and the $p \times p$ matrix having p eigenvalues as its diagonal elements by $\mathbf{D}_p$:

$$\mathbf{A} = (\mathbf{a}_1 \ \ \mathbf{a}_2 \ \ \ldots \ \ \mathbf{a}_p) \quad \text{and} \quad \mathbf{D}_p = \text{diag}(\lambda_1 \ \ \lambda_2 \ \ \ldots \ \ \lambda_p),$$

where $\lambda_1 \geq \lambda_2 \geq \ldots \geq \lambda_p \geq 0$. The relation between the eigenvalues and eigenvectors given by Eq. (2.3) can be expressed by

$$\mathbf{S}\mathbf{A} = \mathbf{A}\mathbf{D}_p, \quad \text{and} \quad \mathbf{A}^\top \mathbf{A} = \mathbf{I}_p,$$

where $\mathbf{I}_p$ is a $p \times p$ identity matrix. We obtain $\mathbf{A} = (\mathbf{a}_1 \ \ \mathbf{a}_2 \ \ \ldots \ \ \mathbf{a}_r)$ by solving

$$\mathbf{S}\mathbf{A} = \mathbf{A}\mathbf{D}_r$$

subject to $\mathbf{A}^\top \mathbf{A} = \mathbf{I}_r$, and then compute $\mathbf{Z} = \mathbf{Y}\mathbf{A}$. Note that

$$\mathbf{Z}^\top \mathbf{Z} = \mathbf{A}^\top \mathbf{Y}^\top \mathbf{Y}\mathbf{A} = n\mathbf{I}_p.$$

2.1.2 Singular Value Decomposition of $\mathbf{Y}$

Let $\mathbf{Y}$ have rank l $(l \leq p)$. From the Eckart-Young decomposition theorem (Eckart and Young 1936), $\mathbf{Y}$ has the following matrix decomposition

$$\mathbf{Y} = \mathbf{U}\mathbf{D}^{1/2}\mathbf{V}^\top, \qquad (2.4)$$

where $\mathbf{U}$, $\mathbf{V}$ and $\mathbf{D}$ have the following properties:

- $\mathbf{U} = (\mathbf{u}_1 \ \mathbf{u}_2 \ \ldots \ \mathbf{u}_l)$ is an $n \times l$ matrix of left singular vectors satisfying $\mathbf{u}_i^\top \mathbf{u}_i = 1$ and $\mathbf{u}_i^\top \mathbf{u}_j = 0$, and $\mathbf{U}^\top \mathbf{U} = \mathbf{I}_l$.
- $\mathbf{V} = (\mathbf{v}_1 \ \mathbf{v}_2 \ \ldots \ \mathbf{v}_l)$ is a $p \times l$ matrix of right singular vectors satisfying $\mathbf{v}_i^\top \mathbf{v}_i = 1$ and $\mathbf{v}_i^\top \mathbf{v}_j = 0$, and $\mathbf{V}^\top \mathbf{V} = \mathbf{I}_l$.
- $\mathbf{D}$ is a $l \times l$ diagonal matrix of eigenvalues of $\mathbf{Y}^\top \mathbf{Y}$ or $\mathbf{Y}\mathbf{Y}^\top$.

We perform spectral decomposition of $\mathbf{Y}^\top \mathbf{Y}$:

$$\mathbf{Y}^\top \mathbf{Y} = \lambda_1 \mathbf{v}_1 \mathbf{v}_1^\top + \lambda_2 \mathbf{v}_2 \mathbf{v}_2^\top + \cdots + \lambda_l \mathbf{v}_l \mathbf{v}_l^\top, \tag{2.5}$$

where $\lambda_1 \geq \lambda_2 \geq \cdots \geq \lambda_l \geq 0$ are eigenvalues of $\mathbf{Y}^\top \mathbf{Y}$ in descending order, and $\mathbf{v}_1, \mathbf{v}_2, \ldots, \mathbf{v}_l$ are the corresponding normalized eigenvalues of length one. The matrices $\mathbf{V}$ and $\mathbf{D}^{1/2}$ based on the decomposition (2.5) are defined as

$$\mathbf{V} = (\mathbf{v}_1 \ \mathbf{v}_2 \ \ldots \ \mathbf{v}_l), \quad \text{and} \quad \mathbf{D}^{1/2} = \mathrm{diag}(\sqrt{\lambda_1} \ \sqrt{\lambda_2} \ \ldots \ \sqrt{\lambda_l}).$$

From Eq. (2.4), we have

$$\mathbf{Z} = \mathbf{Z}\mathbf{A}^\top \mathbf{A} = \mathbf{Y}\mathbf{A} = \mathbf{U}\mathbf{D}^{1/2}.$$

Then the matrix $\mathbf{U}$ under restrictions $\mathbf{u}_i^\top \mathbf{u}_i = 1$ and $\mathbf{u}_i^\top \mathbf{u}_j = 0$ is given by

$$\mathbf{U} = \left(\frac{1}{\sqrt{\lambda_1}} \mathbf{Y}\mathbf{v}_1 \ \ \frac{1}{\sqrt{\lambda_2}} \mathbf{Y}\mathbf{v}_2 \ \ \cdots \ \ \frac{1}{\sqrt{\lambda_l}} \mathbf{Y}\mathbf{v}_l \right).$$

2.2 Quantification of Qualitative Data

Optimal scaling is a quantification technique that optimally assigns numerical values to qualitative scales within the restrictions of the measurement characteristics of the qualitative variables (Young 1981).

Let $\mathbf{y}_j$ of $\mathbf{Y}$ be a qualitative vector with K_j categories. To quantify $\mathbf{y}_j$, the vector is coded by using an $n \times K_j$ indicator matrix

$$\mathbf{G}_j = (g_{jik}) = \begin{pmatrix} g_{j11} & \cdots & g_{j1K_j} \\ \vdots & \vdots & \vdots \\ g_{jn1} & \cdots & g_{jnK_j} \end{pmatrix} = (\mathbf{g}_{j1} \ \cdots \ \mathbf{g}_{jK_j}),$$

where

$$g_{jik} = \begin{cases} 1 & \text{if object } i \text{ belongs to category } k, \\ 0 & \text{if object } i \text{ belongs to some other category } k'(\neq k). \end{cases}$$

For example, given

$$
\mathbf{Y} = (\mathbf{y}_1 \ \mathbf{y}_2 \ \mathbf{y}_3) = \begin{pmatrix} \text{Blue} & \text{Yes} & 4 \\ \text{Red} & \text{No} & 3 \\ \text{Green} & \text{Yes} & 1 \\ \text{Green} & \text{No} & 2 \\ \text{Blue} & \text{Yes} & 1 \end{pmatrix},
$$

the indicator matrix of $\mathbf{Y}$ is

$$
\mathbf{G} = (\mathbf{G}_1 \ \mathbf{G}_2 \ \mathbf{G}_3) = \begin{pmatrix} 0\,0\,1 & 1\,0 & 0\,0\,0\,1 \\ 1\,0\,0 & 0\,1 & 0\,0\,1\,0 \\ 0\,1\,0 & 1\,0 & 1\,0\,0\,0 \\ 0\,1\,0 & 0\,1 & 0\,1\,0\,0 \\ 0\,0\,1 & 1\,0 & 1\,0\,0\,0 \end{pmatrix}.
$$

Thus we have

$$
\mathbf{y}_1 = \mathbf{G}_1 \begin{pmatrix} \text{Red} \\ \text{Green} \\ \text{Blue} \end{pmatrix}, \quad \mathbf{y}_2 = \mathbf{G}_2 \begin{pmatrix} \text{Yes} \\ \text{No} \end{pmatrix}, \quad \mathbf{y}_3 = \mathbf{G}_3 \begin{pmatrix} 1 \\ 2 \\ 3 \\ 4 \end{pmatrix}.
$$

Optimal scaling finds $K_j \times 1$ category quantifications $\mathbf{q}_j$ under the restrictions imposed by the measurement level of variable j and transforms $\mathbf{y}_j$ into an optimally scaled vector $\mathbf{y}_j^* = \mathbf{G}_j \mathbf{q}_j$. There are different ways for quantifying observed data of nominal, ordinal and numerical variables:

- Nominal scale data: The quantification is unrestricted. Objects i and $h (\neq i)$ in the same category for variable j obtain the same quantification. Thus, if $y_{ji} = y_{jh}$ then $y_{ji}^* = y_{jh}^*$.
- Ordinal scale data: The quantification is restricted to the order of categories. If observed categories y_{ji} and y_{jh} for objects i and h in variable j have order $y_{ji} > y_{jh}$ then quantified categories have order $y_{ji}^* \geq y_{jh}^*$.
- Numerical data: The observed vector $\mathbf{y}_j$ for variable j replaces $\mathbf{y}_j^*$ by standardizing with zero mean and unit variance.

2.3 Nonlinear PCA

PCA assumes that data are quantitative and thus it is not directly applicable to qualitative data such as nominal and ordinal data. When PCA handles mixed quantitative and qualitative data, the qualitative data must be quantified. In nonlinear PCA, the qualitative data of nominal and ordinal variables are nonlinearly transformed into quantitative data. Thus, PCA with optimal scaling is called nonlinear PCA (Gifi 1990).

Nonlinear PCA reveals nonlinear relationships among variables with different measurement levels and therefore presents a more flexible alternative to ordinary PCA. Nonlinear PCA can find solutions by minimizing two types of loss functions; a low-rank approximation of $\mathbf{Y}^*$ extended to Eq. (2.2) and homogeneity analysis with restrictions. We show the loss functions and provide the ALS algorithm used for minimizing these loss functions.

2.3.1 Low-Rank Matrix Approximation

In the presence of qualitative variables in $\mathbf{Y}$, the loss function (2.2) is expressed as

$$\sigma_L(\mathbf{Z}, \mathbf{A}, \mathbf{Y}^*) = \text{tr}(\mathbf{Y}^* - \hat{\mathbf{Y}})^\top (\mathbf{Y}^* - \hat{\mathbf{Y}}) = \text{tr}(\mathbf{Y}^* - \mathbf{Z}\mathbf{A}^\top)^\top (\mathbf{Y}^* - \mathbf{Z}\mathbf{A}^\top) \quad (2.6)$$

and is minimized over $\mathbf{Z}$, $\mathbf{A}$ and $\mathbf{Y}^*$ under the restrictions

$$\mathbf{Y}^{*\top}\mathbf{1}_n = \mathbf{0}_p \quad \text{and} \quad \text{diag}\left[\frac{\mathbf{Y}^{*\top}\mathbf{Y}^*}{n}\right] = \mathbf{I}_p, \quad (2.7)$$

where $\mathbf{1}_n$ and $\mathbf{0}_p$ are vectors of ones and zeros of length n and p, respectively. Optimal scaling for $\mathbf{Y}^*$ can be performed separately and independently for each variable, and then the loss function (2.6) can be rewritten as

$$\sigma_L(\mathbf{Z}, \mathbf{A}, \mathbf{Y}^*) = \sum_{j=1}^p (\mathbf{y}_j^* - \mathbf{Z}\mathbf{a}_j^\top)^\top (\mathbf{y}_j^* - \mathbf{Z}\mathbf{a}_j^\top) = \sum_{j=1}^p \sigma_{Lj}(\mathbf{Z}, \mathbf{a}_j, \mathbf{y}_j^*). \quad (2.8)$$

when minimizing independently each $\sigma_{Lj}(\mathbf{Z}, \mathbf{a}_j, \mathbf{y}_j^*)$ under the measurement restrictions on variable j, we can minimize $\sigma(\mathbf{Z}, \mathbf{A}, \mathbf{Y}^*)$.

2.3.2 Homogeneity Analysis

Homogeneity analysis maximizes the homogeneity of several categorical variables and quantifies the categories of each variable such that the homogeneity is maximized (Gifi 1990). Let $\mathbf{Z}$ be $n \times r$ object scores (component scores) and $\mathbf{W}_j$ be $K_j \times r$ category quantifications of variable j ($j = 1, \ldots, p$). The loss function measuring the departure from homogeneity is given by

$$\sigma_H(\mathbf{Z}, \mathbf{W}) = \sum_{j=1}^{p} \mathrm{tr}(\mathbf{Z} - \mathbf{G}_j\mathbf{W}_j)^\top(\mathbf{Z} - \mathbf{G}_j\mathbf{W}_j)$$

$$= \sum_{j=1}^{p} \sigma_{Hj}(\mathbf{Z}, \mathbf{W}_j) \tag{2.9}$$

and is minimized over $\mathbf{Z}$ and $\mathbf{W}$ under the restrictions

$$\mathbf{Z}^\top\mathbf{1}_n = \mathbf{0}_r, \quad \text{and} \quad \mathbf{Z}^\top\mathbf{Z} = n\mathbf{I}_r. \tag{2.10}$$

The minimum of $\sigma_H(\mathbf{Z}, \mathbf{W})$ is obtained by separately minimizing each $\sigma_{Hj}(\mathbf{Z}, \mathbf{W}_j)$.

Gifi (1990) defines nonlinear PCA as homogeneity analysis imposing a *rank-one restriction* whose form is

$$\mathbf{W}_j = \mathbf{q}_j\mathbf{a}_j, \tag{2.11}$$

where $\mathbf{q}_j$ is a $K_j \times 1$ vector of category quantifications and $\mathbf{a}_j$ is a $1 \times r$ vector of weights (component loadings). Nominal variables on which restriction (2.11) is imposed are called *single nominal variables* and variables without restrictions are *multiple nominal variables*.

To minimize $\sigma_{Hj}(\mathbf{Z}, \mathbf{W}_j)$ under restriction (2.11), we first obtain the least squares estimate $\tilde{\mathbf{W}}_j$ of $\mathbf{W}_j$. For a fixed $\tilde{\mathbf{W}}_j$, $\sigma_{H_j}(\mathbf{Z}, \mathbf{W}_j)$ can be partitioned as

$$\sigma_{H_j}(\mathbf{Z}, \mathbf{W}_j) = \mathrm{tr}(\mathbf{Z} - \mathbf{G}_j\mathbf{W}_j)^\top(\mathbf{Z} - \mathbf{G}_j\mathbf{W}_j)$$
$$= \mathrm{tr}(\mathbf{Z} - \mathbf{G}_j\tilde{\mathbf{W}})^\top(\mathbf{Z} - \mathbf{G}_j\tilde{\mathbf{W}}_j)$$
$$+ \mathrm{tr}(\mathbf{q}_j\mathbf{a}_j - \tilde{\mathbf{W}}_j)^\top(\mathbf{G}_j^\top\mathbf{G}_j)(\mathbf{q}_j\mathbf{a}_j - \tilde{\mathbf{W}}_j). \tag{2.12}$$

We then minimize the second term on the right hand side of Eq. (2.12) over $\mathbf{q}_j$ and $\mathbf{a}_j$ under the restrictions imposed by the measurement level of variable j.

Each column vector of $\mathbf{Y}^*$ under restriction (2.11) is computed by

$$\mathbf{y}_j^* = \mathbf{G}_j\mathbf{q}_j.$$

Then Eq. (2.9) under restriction (2.10) is expressed as

$$\sigma_H(\mathbf{Z}, \mathbf{W}) = \sum_{i=1}^{p} \mathrm{tr}(\mathbf{Z} - \mathbf{G}_j\mathbf{W}_j)^\top(\mathbf{Z} - \mathbf{G}_j\mathbf{W}_j)$$

$$= np - 2\sum_{i=1}^{p} \mathrm{tr}(\mathbf{a}_j^\top\mathbf{y}_j^{*\top}\mathbf{Z}_j) + \sum_{i=1}^{p} \mathrm{tr}(\mathbf{a}_j^\top\mathbf{y}_j^{*\top}\mathbf{y}_j^*\mathbf{a}_j)$$

$$= np - 2\mathrm{tr}(\mathbf{A}^\top\mathbf{Y}^{*\top}\mathbf{Z}) + \mathrm{tr}(\mathbf{A}^\top\mathbf{A}).$$

when expanding Eq. (2.6) under restriction (2.7), we also obtain

$$\sigma_L(\mathbf{Z}, \mathbf{A}, \mathbf{Y}^*) = \mathrm{tr}(\mathbf{Y}^* - \mathbf{Z}\mathbf{A}^\top)^\top (\mathbf{Y}^* - \mathbf{Z}\mathbf{A}^\top)^\top$$
$$= np - 2\mathrm{tr}(\mathbf{A}^\top \mathbf{Y}^{*\top} \mathbf{Z}) + \mathrm{tr}(\mathbf{A}\mathbf{A}^\top).$$

Thus, minimizing the loss function (2.9) is equivalent to minimizing the loss function (2.6) under restrictions (2.7) and (2.10).

2.3.3 Alternating Least Squares Algorithm for Nonlinear PCA

The minimization of loss functions (2.6) and (2.9) has to take place with respect to both parameters of $\mathbf{Y}^*$ and $(\mathbf{Z}, \mathbf{A})$ and both of $\mathbf{Z}$ and $\mathbf{W}$, although we can not find simultaneously the solutions of these parameters. The alternating least squares (ALS) algorithm is utilized to solve such minimization problem.

We describe the general procedure of the ALS algorithm. Let $\sigma(\theta_1, \theta_2)$ be a loss function and (θ_1, θ_2) be the parameter matrices of the function. We denote the t-th estimate of θ as $\theta^{(t)}$. To minimize $\sigma(\theta_1, \theta_2)$ over θ_1 and θ_2, the ALS algorithm updates the estimates of θ_1 and θ_2 by solving the least squares problem for each parameter:

$$\theta_1^{(t+1)} = \arg\min_{\theta_1} \sigma(\theta_1, \theta_2^{(t)}),$$
$$\theta_2^{(t+1)} = \arg\min_{\theta_2} \sigma(\theta_1^{(t+1)}, \theta_2).$$

If each update of the ALS algorithm improves the value of the loss function and if the function is bounded, the function will be locally minimized over the entire set of parameters (Krijnen 2006).

We show the two ALS algorithms typically employed in nonlinear PCA; PRINCIPALS (Young et al. 1978) and PRINCALS (Gifi 1990).

2.3.3.1 PRINCIPALS

PRINCIPALS developed by Young et al. (1978) is the ALS algorithm that minimizes the loss function (2.8). PRINCIPALS accepts single nominal, ordinal and numerical variables, and alternates between two estimation steps. The first step estimates the model parameters $\mathbf{Z}$ and $\mathbf{A}$ for ordinary PCA, and the second obtains the estimate of the data parameter $\mathbf{Y}^*$ for optimally scaled data.

For the initialization of PRINCIPALS, the initial data $\mathbf{Y}^{*(0)}$ are determined under the measurement restrictions for each variable and are then standardized to satisfy restriction (2.7). The observed data $\mathbf{Y}$ may be used as $\mathbf{Y}^{*(0)}$ after standardizing each column of $\mathbf{Y}$ under restriction (2.7). Given the initial data $\mathbf{Y}^{*(0)}$, PRINCIPALS iterates the following two steps:

- *Model estimation step*: By solving the eigen-decomposition of $\mathbf{Y}^{*(t)\top}\mathbf{Y}^{*(t)}/n$ or the singular value decomposition of $\mathbf{Y}^{*(t)}$, obtain $\mathbf{A}^{(t+1)}$ and compute $\mathbf{Z}^{(t+1)} = \mathbf{Y}^{*(t)}\mathbf{A}^{(t+1)}$. Update $\hat{\mathbf{Y}}^{(t+1)} = \mathbf{Z}^{(t+1)}\mathbf{A}^{(t+1)\top}$.
- *Optimal scaling step*: Obtain $\mathbf{Y}^{*(t+1)}$ by separately estimating $\mathbf{y}_j^*$ for each variable j. Compute $\mathbf{q}_j^{(t+1)}$ for nominal variables as

$$\mathbf{q}_j^{(t+1)} = (\mathbf{G}_j^\top \mathbf{G}_j)^{-1}\mathbf{G}_j^\top \hat{\mathbf{y}}_j^{(t+1)}.$$

Re-compute $\mathbf{q}_j^{(t+1)}$ for ordinal variables using the monotone regression (Kruskal 1964). For nominal and ordinal variables, update $\mathbf{y}_j^{*(t+1)} = \mathbf{G}_j\mathbf{q}_j^{(t+1)}$ and stan-

Table 2.1 Sleeping bag data from Prediger (1997)

	Temperature	Weight	Price	Material	Quality rate
One kilo bag	7	940	149	Liteloft	3
Sund	3	1880	139	Hollow ber	1
Kompakt basic	0	1280	249	MTI Loft	3
Finmark tour	0	1750	179	Hollow ber	1
Interlight Lyx	0	1900	239	Thermolite	1
Kompakt	−3	1490	299	MTI Loft	2
Touch the cloud	−3	1550	299	Liteloft	2
Cat's meow	−7	1450	339	Polarguard	3
Igloo super	−7	2060	279	Terraloft	1
Donna	−7	1850	349	MTI Loft	2
Tyin	−15	2100	399	Ultraloft	2
Travellers dream	3	970	379	Goose-downs	3
Yeti light	3	800	349	Goose-downs	3
Climber	−3	1690	329	Duck-downs	2
Viking	−3	1200	369	Goose-downs	3
Eiger	−3	1500	419	Goose-downs	2
Climber light	−7	1380	349	Goose-downs	3
Cobra	−7	1460	449	Duck-downs	3
Cobra comfort	−10	1820	549	Duck-downs	2
Fox re	−10	1390	669	Goose-downs	3
Mont Blanc	−15	1800	549	Goose-downs	3

dardize $\mathbf{y}_j^{*(t+1)}$. For numerical variables, standardize observed vector $\mathbf{y}_j$ and set $\mathbf{y}_j^{*(t+1)} = \mathbf{y}_j$.

2.3.3.2 PRINCALS

PRINCALS is the ALS algorithm developed by Gifi (1990) and can handle multiple nominal variables in addition to the single nominal, ordinal and numerical variables. We denote the set of multiple variables by $\mathbf{J}_M$ and the set of single variables having single nominal and ordinal scales and numerical measurements by $\mathbf{J}_S$. From Eqs. (2.9) and (2.12), the loss function to be minimized by PRINCALS is given by

$$\sigma_H(\mathbf{Z}, \mathbf{W}) = \sum_{j \in \mathbf{J}_M} \sigma_{Hj}(\mathbf{Z}, \mathbf{W}_j) + \sum_{j \in \mathbf{J}_S} \sigma_{Hj}(\mathbf{Z}, \mathbf{W}_j).$$

For the initialization of PRINCALS, we determine the initial values of $\mathbf{Z}$ and $\mathbf{W}$. The matrix $\mathbf{Z}^{(0)}$ is initialized with random numbers under restriction (2.10), and $\mathbf{W}_j^{(0)}$ is obtained as $\mathbf{W}_j^{(0)} = (\mathbf{G}_j^\top \mathbf{G}_j)^{-1} \mathbf{G}_j^\top \mathbf{Z}^{(0)}$. For each variable $j \in \mathbf{J}_S$, $\mathbf{q}_j^{(0)}$ is defined as the first K_j successive integers under the normalization restriction. The vector $\mathbf{a}_j$ is initialized as $\mathbf{a}_j^{(0)} = \mathbf{Z}^{(0)\top} \mathbf{G}_j \mathbf{q}_j^{(0)}$, and rescaled to unit length. Given these initial values, PRINCALS iterates the following steps (Michailidis and de Leeuw 1998):

- *Estimation of category quantifications:* Compute $\mathbf{W}_j^{(t+1)}$ for $j = 1, \ldots, p$ as

$$\mathbf{W}_j^{(t+1)} = (\mathbf{G}_j^\top \mathbf{G}_j)^{-1} \mathbf{G}_j^\top \mathbf{Z}^{(t)}.$$

Table 2.2 Quantification of Material and Quality rate

Material	
Duck-downs	−0.168
Goose-downs	−0.147
Hollow ber	0.473
Liteloft	0.046
MTI loft	0.014
Polarguard	−0.005
Terraloft	0.313
Thermolite	0.390
Ultraloft	−0.250
Quality rate	
1	−0.450
2	0.106
3	0.106

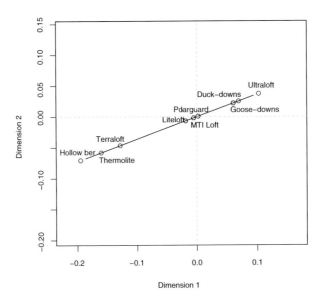

Fig. 2.1 Category plot for material

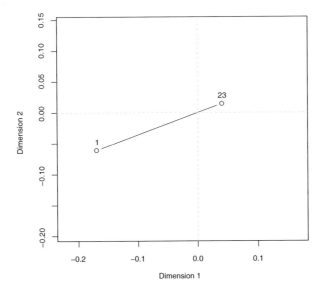

Fig. 2.2 Category plot for Quality rate

Table 2.3 Optimal scaled sleeping bag data

	Temperature	Weight	Price	Material	Quality rate
One kilo bag	0.425	−0.324	−0.342	0.046	0.106
Sund	0.290	0.252	−0.388	0.473	−0.450
Kompakt basic	0.155	−0.216	−0.202	0.014	0.106
Finmark tour	0.155	0.108	−0.295	0.473	−0.450
Interlight Lyx	0.155	0.288	−0.248	0.390	−0.450
Kompakt	0.019	−0.036	−0.109	0.014	0.106
Touch the cloud	0.019	0.036	−0.109	0.046	0.106
Cat's meow	−0.116	−0.108	−0.016	−0.005	0.106
Igloo super	−0.116	0.324	−0.155	0.313	−0.450
Donna	−0.116	0.216	0.031	0.014	0.106
Tyin	−0.387	0.360	0.171	−0.250	0.106
Travellers dream	0.290	−0.288	0.124	−0.147	0.106
Yeti light	0.290	−0.360	0.031	−0.147	0.106
Climber	0.019	0.072	−0.062	−0.168	0.106
Viking	0.019	−0.252	0.078	−0.147	0.106
Eiger	0.019	0.000	0.217	−0.147	0.106
Climber light	−0.116	−0.180	0.031	−0.147	0.106
Cobra	−0.116	−0.072	0.264	−0.168	0.106
Cobra comfort	−0.251	0.180	0.311	−0.168	0.106
Fox re	−0.251	−0.144	0.357	−0.147	0.106
Mont Blanc	−0.387	0.144	0.311	−0.147	0.106

- For the multiple variables in $\mathbf{J}_M$, set $\mathbf{W}_j^{(t+1)}$ to the estimate of multiple category quantifications.
- For the single variables in $\mathbf{J}_S$, update $\mathbf{a}_j^{(t+1)}$ by

$$\mathbf{a}_j^{(t+1)\top} = \mathbf{W}_j^{(t+1)\top}(\mathbf{G}_j^\top \mathbf{G}_j)\mathbf{q}_j^{(t)} \Big/ \mathbf{q}_j^{(t)\top}(\mathbf{G}_j^\top \mathbf{G}_j)\mathbf{q}_j^{(t)}$$

and compute $\mathbf{q}_j^{(t+1)}$ for nominal variables by

$$\mathbf{q}_j^{(t+1)} = \mathbf{W}_j^{(t+1)}\mathbf{a}_j^{(t+1)\top} \Big/ \mathbf{a}_j^{(t+1)}\mathbf{a}_j^{(t+1)\top}.$$

Re-compute $\mathbf{q}_j^{(t+1)}$ for ordinal variables using the monotone regression in a similar manner as for PRINCIPALS. For numerical variables, standardize observed

Table 2.4 Component scores

	Z_1	Z_2
One kilo bag	−0.104	0.185
Sund	−0.208	−0.048
Kompakt basic	−0.010	−0.084
Finmark tour	−0.172	−0.120
Interlight Lyx	−0.167	−0.040
Kompakt	0.019	−0.065
Touch the cloud	0.010	−0.044
Cat's meow	0.024	−0.059
Igloo super	−0.132	−0.036
Donna	0.006	0.056
Tyin	0.078	0.105
Travellers dream	−0.006	0.224
Yeti light	−0.014	0.194
Climber	0.042	0.003
Viking	0.078	−0.109
Eiger	0.080	−0.013
Climber light	0.053	−0.061
Cobra	0.079	−0.008
Cobra comfort	0.111	0.010
Fox re	0.136	−0.107
Mont Blanc	0.097	0.0182

Table 2.5 Factor loadings

	Temperature	Weight	Price	Material	Quality rate
Z_1	−0.330	−0.204	0.396	−0.411	0.377
Z_2	0.179	0.348	0.045	−0.149	0.136

vector $\mathbf{y}_j$ and compute $\mathbf{q}_j^{(t+1)} = (\mathbf{G}_j^\top \mathbf{G}_j)^{-1} \mathbf{G}_j^\top \mathbf{y}_j$. Update $\mathbf{W}_j^{(t+1)} = \mathbf{q}_j^{(t+1)} \mathbf{a}_j^{(t+1)}$ for ordinal and numerical variables.

- *Update of object scores:* Compute $\mathbf{Z}^{(t+1)}$ by

$$\mathbf{Z}^{(t+1)} = \frac{1}{p} \sum_{j=1}^{p} \mathbf{G}_j \mathbf{W}_j^{(t+1)}.$$

Column-wise center and orthonormalize $\mathbf{Z}^{(t+1)}$.

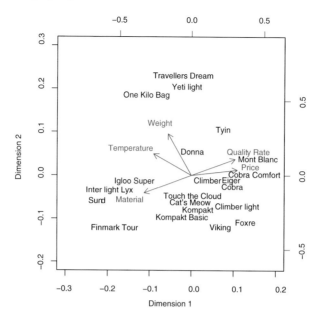

Fig. 2.3 Biplot of the first two principal components

2.4 Example: Sleeping Bags

We illustrate nonlinear PCA using sleeping bag data from Prediger (1997) given in Table 2.1. The data were collected on 21 sleeping bags with Temperature, Weight, Price, Material and Quality Rate. Quality Rate is scaled from 1 to 3 such that the higher value is the better one. The first three variables are numerical, Material is nominal and Quality Rate is ordinal. The computation for quantifying qualitative data and PCA is performed by the R package `homals` of De Leeuw and Mair (2009) that provides the ALS algorithm for homogeneity analysis. When imposing the rank-one restrictions on Material and Quality Rate, `homals` is the same as PRINCALS. We set $r = 2$ and obtain the following results.

Table 2.2 reports the quantified values of Material and Quality Rate. Then Material are quantified without order restriction due to the nominal variable, while the quantification of Quality Rate is restricted to the order of categories. Figures 2.1 and 2.2 are the plots of the category quantifications of Material and Quality Rate. These figures graphically show the order restrictions for these variables. Table 2.3 shows optimal scaled sleeping bag data. The component scores and factor loadings are given in Tables 2.4 and 2.5, respectively. Figure 2.3 is the biplot of the first two principal components. We can interpret the data using ordinary PCA.

References

De Leeuw, J., Mair, P.: A general framework for multivariate analysis with optimal scaling: the R package homals. J. Stat. Softw. **31**, 1–21 (2009)

Eckart, C., Young, G.: The approximation of one matrix by another of lower rank. Psychometrika **1**, 211–218 (1936)

Gifi, A.: Nonlinear Multivariate Analysis. Wiley, Chichester (1990)

Krijnen, W.P.: Convergence of the sequence of parameters generated by alternating least squares algorithms. Comput. Stat. Data Anal. **51**, 481–489 (2006)

Kruskal, J.B.: Nonmetric multidimensional scaling: a numerical method. Psychometrika **29**, 115–129 (1964)

Michailidis, G., de Leeuw, J.: The Gifi system of descriptive multivariate analysis. Stat. Sci. **13**, 307–336 (1998)

Prediger, S.: Symbolic objects in formal concept analysis. In: Mineau, G., Fall, A (eds.) Proceedings of the 2nd International Symposium on Knowledge, Retrieval, Use, and Storage for Efficiency (1997)

Young, F.W.: Quantitative analysis of qualitative data. Psychometrika **46**, 357–388 (1981)

Young, F.W., Takane, Y., de Leeuw, J.: Principal components of mixed measurement level multivariate data: an alternating least squares method with optimal scaling features. Psychometrika **43**, 279–281 (1978)

Chapter 3
Multiple Correspondence Analysis

Abstract Multiple correspondence analysis (MCA) is a widely used technique to analyze categorical data and aims to reduce large sets of variables into smaller sets of components that summarize the information contained in the data. The purpose of MCA is the same as that of principal component analysis (PCA), and MCA can be regarded as an adaptation to the categorical data of PCA (Jolliffe, Principal Component Analysis, 2002). There are various approaches to formulate an MCA. We introduce a formulation in which the quantified data matrix is approximated by a lower-rank matrix using the quantification technique proposed by Murakami et al. (Non-metric principal component analysis for categorical variables with multiple quantifications, 1999).

Keywords Quantification · Low-rank approximation · Rank-restriction of quantification parameters · Orthogonal procrustes analysis

3.1 Introduction

In social, psychological, or behavioral studies, researchers frequently encounter a large number of multivariate categorical variables. For example, in questionnaires, participants are asked to choose one of several response alternatives for each set of questions. Multiple correspondence analysis (MCA) is a widely used technique for analyzing such categorical data and aims to reduce large sets of variables into smaller sets of components that summarize the information contained in the data. The purpose of MCA is the same as that of principal component analysis (PCA), and MCA can be regarded as an adaptation to categorical data of PCA (Jolliffe 2002). There are various approaches by which to formulate MCA (e.g., Benzecri 1973, 1992; Gifi 1990; Hayashi 1952; Nishisato 2006), although these approaches have been shown to give essentially equivalent solutions originating in different theoretical foundations (Tenenhaus and Young 1985).

Similar to PCA, MCA can also be formulated as approximating a data matrix by a lower-rank matrix using the quantification technique (Murakami et al. 1999; Murakami 1999; Adachi and Murakami 2011). Quantification is a widely used statistical technique for analyzing categorical variables, whereby the observed categorical variables are transformed into quantitative scores such that the data analysis model most closely matches the observations (Gifi 1990; Young 1981).

3.2 Formulation

If K_i denotes the number of categories in the i-th variable and G_i denotes the $n \times K_i$ indicator matrix, then $G_i Q_i$ represents the multiple quantification of i-th variable where Q_i (K_i-categories $\times$ R_i-dimensions) is the quantification parameter matrix of the i-th variable. When the i-th variable is categorical, the observed variable is transformed into quantitative scores by a quantification procedure. The loss function is expressed as

$$\sum_{i=1}^{p} \| G_i Q_i - Z A_i^\top \|^2, \tag{3.1}$$

is minimized over Z, A_i, and Q_i ($i = 1, \ldots, p$), and subject to the following constraints

$$n^{-1} Z^\top Z = I_r, \qquad\qquad J Z = Z, \tag{3.2}$$

$$1_n^\top G_i Q_i = 0_{K_i}^\top, \quad n^{-1} Q_i^\top G_i^\top G_i Q_i = I_{R_i} \tag{3.3}$$

where J ($n \times n$) is a centering matrix, Z ($n \times r$) is a component score matrix, and A_i ($R_i \times r$) is a loading matrix, which are the corresponding model parameters of the multidimensionally quantified i-th variable, 1_n is an n-dimensional vector whose elements are all ones and 0_{K_i} is a K_i-dimensional vector whose elements are all zeros. The component score matrix and the quantified data are assumed to be centered and orthonormal in a column-wise manner.

The optimal solution is given by the singular value decomposition (SVD) of indicator matrices of the observed variables. Denote $G = [G_1, \ldots, G_p]$, $D_i = G_i^\top G_i$, $D = \text{b-diag}(D_1, \ldots, D_p)$, $W_i = Q_i A_i$, $W = [W_1, \ldots, W_p]$. W is called a category score matrix in MCA. Here we define b-diag($\bullet$) as the operator of the block diagonal matrix. Then, the loss function (3.1) can be rewritten as

$$\| J G D^{-1/2} - Z W^\top D^{1/2} \|^2. \tag{3.4}$$

Let the SVD of $J G D^{-1/2} = M \Theta N^\top$. The optimal solutions can be obtained as $Z = M_r$ and $W = D^{-1/2} N_r \Theta_r$. The category score matrix is also expressed as $W = D^{-1} G^\top J Z$. The rank of W_i is at most the smaller of the number of components or categories minus one, because $rank(\mathbf{JG}_i\mathbf{D}_i^{-1}) = K_i - 1$ and $rank(\mathbf{Z}) = r$. Adachi and Murakami (2011) have shown that the category score matrix $\mathbf{W}$ can be decomposed into the product of the quantification matrix and the loading matrix. Let us denote the SVD of $\mathbf{W}_i = \mathbf{U}_i\,\Xi_i\mathbf{V}_i^\top$. We can obtain the optimal solutions

$$\mathbf{Q}_i = \mathbf{U}_i\mathbf{\Phi}_i, \tag{3.5}$$
$$\mathbf{A}_i = \mathbf{\Phi}_i^{-1}\Xi_i\mathbf{V}_i^\top, \tag{3.6}$$

if the nonsingular matrix $\mathbf{\Phi}_i$ $(i = 1, \ldots, p)$ that are satisfied with $n^{-1}\mathbf{\Phi}_i^\top\mathbf{U}_i^\top\mathbf{D}_i\mathbf{U}_i$ $\mathbf{\Phi}_i = \mathbf{I}_{R_i}$ exist. $\mathbf{\Phi}_i$ can be obtained by the eigenvalue decomposition of $n^{-1}\mathbf{U}_i^\top\mathbf{D}_i\mathbf{U}_i = \mathbf{P}_i\mathbf{\Psi}_i^2\mathbf{P}_i^\top$ where $\mathbf{P}_i$ is the eigenvector matrix and $\mathbf{\Psi}_i$ is the diagonal matrix whose diagonal elements are eigenvalues. Then, we have $\mathbf{\Phi}_i = \mathbf{P}_i\mathbf{\Psi}^{-1}$. Substituting $\mathbf{\Phi}_i$ into the optimal solutions (3.5) and (3.6), the quantification parameter matrix and the loading matrix are expressed as $\mathbf{Q}_i = \mathbf{U}_i\mathbf{P}_i\mathbf{\Psi}^{-1}$ and $\mathbf{A}_i = \mathbf{\Psi}\mathbf{P}_i^\top\Xi_i\mathbf{V}_i^\top$, respectively.

3.3 Rank-Restricted MCA

In ordinary MCA, the number of quantification dimensions is fixed to $\min(K_i - 1, r)$ for each variable, and the parameters can be uniquely determined. When the number of dimensions is set in $1 \leq R_i < \min(K_i - 1, r)$, the solutions cannot be uniquely determined and thus the parameters can be estimated by alternate iterations of the operations that quantify the categorical variables and that obtain component score and loading matrices. This type of MCA is referred to as rank-restricted MCA (Murakami et al. 1999), which includes ordinary MCA as a special case. When all categorical variables are unidimensionally quantified, the quantification parameters are assumed to be $\mathbf{1}_n^\top\mathbf{G}_i\mathbf{q}_i = \mathbf{0}_{K_i}^\top, n^{-1}\mathbf{q}_i^\top\mathbf{G}_i^\top\mathbf{G}_i\mathbf{q}_i = 1$. Note that the order-restrictions are imposed to the quantification parameters as well as the rank-one restrictions in the ordinal variables case, and that rank-restricted MCA reduces to nonlinear PCA as mentioned in Chap. 2. In the case of numerical variables, standardized numerical variables also satisfy the constraints above. In other words, standardization of numerical variables is a restricted version of a single quantification ($R_i = 1$), in which the quantification parameters are already known, and nonlinear PCA is equivalent to PCA when all variables are numerical. The hierarchical relationships between MCA, NPCA, and PCA are summarized in Table 3.1.

Table 3.1 Relationships between MCA, nonlinear PCA, and PCA

Quantification	PCA model
Non	PCA
Single	Nonlinear PCA
Multiple	MCA

3.4 Algorithm of Rank-Restricted MCA

We introduce an alternating least squares algorithm for rank-restricted MCA is described as follow.

We reparametrize $\mathbf{G}_i\mathbf{Q}_i$ as

$$\mathbf{G}_i\mathbf{Q}_i = \mathbf{G}_i(\mathbf{G}_i^\top\mathbf{G}_i)^{-1/2}(\mathbf{G}_i^\top\mathbf{G}_i)^{1/2}\mathbf{Q}_i = \mathbf{\Gamma}_i\mathbf{\Lambda}_i,$$

where $\mathbf{\Gamma}_i = n^{1/2}\mathbf{G}_i(\mathbf{G}_i^\top\mathbf{G}_i)^{-1/2}$ and $\mathbf{\Lambda}_i = n^{-1/2}(\mathbf{G}_i^\top\mathbf{G}_i)^{1/2}\mathbf{Q}_i$. Substituting $\mathbf{\Gamma}_i$ and $\mathbf{\Lambda}_i$ in the loss function (3.1), the problem reduces to minimizing

$$\|\mathbf{\Gamma}_i\mathbf{\Lambda}_i - \mathbf{Z}\mathbf{A}_i^\top\|^2, \tag{3.7}$$

over each $\mathbf{\Lambda}_i$, subject to the constraints (3.3) for fixed model parameters. This minimization can be viewed as an orthogonal Procrustes problem because $\mathbf{\Lambda}_i$ is column-orthonormal matrix where $\mathbf{\Lambda}_i^\top\mathbf{\Lambda}_i = n^{-1}\mathbf{Q}_i^\top\mathbf{G}_i^\top\mathbf{G}_i\mathbf{Q}_i = \mathbf{I}_{R_i}$. Let us define the SVD as $\mathbf{\Gamma}_i^\top\mathbf{Z}\mathbf{A}_i^\top = \mathbf{K}\mathbf{\Delta}\mathbf{L}^\top$ with $\mathbf{K}^\top\mathbf{K} = \mathbf{L}^\top\mathbf{L} = \mathbf{L}\mathbf{L}^\top = \mathbf{I}_{R_i}$, and let $\mathbf{\Delta}$ be the $R_i \times R_i$ diagonal matrix whose diagonal elements are arranged in descending order. The optimal $\mathbf{\Lambda}_i$ is given by

$$\mathbf{\Lambda}_i = \mathbf{K}\mathbf{L}^\top \tag{3.8}$$

(ten Berge 1993). Then we have

$$\mathbf{Q}_i = n^{1/2}(\mathbf{G}_i^\top\mathbf{G}_i)^{-1/2}\mathbf{K}\mathbf{L}^\top. \tag{3.9}$$

The rank of $\mathbf{\Gamma}_i^\top\mathbf{Z}\mathbf{A}_i^\top$ is at most equal to the smaller of $K_i - 1$ or r. Consequently, the dimension of quantification R_i is upper bounded by $\min(K_i - 1, r)$. The orthogonal Procrustes problems expressed above are solved for each variable.

Next, we consider the optimization of $\mathbf{Z}$. Recall the loss function (3.1)

$$\sum_{i=1}^{p} \|\mathbf{G}_i\mathbf{Q}_i - \mathbf{Z}\mathbf{A}_i^\top\|^2.$$

This loss function is minimized over $\mathbf{Z}$ subject to $n^{-1}\mathbf{Z}^\top\mathbf{Z} = \mathbf{I}_r$ for given $\mathbf{Q}_i$ and $\mathbf{A}_i$. This minimization can also be viewed as an orthogonal Procrustes problem and the optimal solution of $\mathbf{Z}$ is given by the SVD of $\sum \mathbf{G}_i\mathbf{Q}_i\mathbf{A}_i$.

After updating the component score matrix $\mathbf{Z}$, the quantification matrix $\mathbf{Q}_i\mathbf{A}_i$ is obtained by solving the regression problems

$$\mathbf{A}_i = n^{-1}\mathbf{Q}_i^\top\mathbf{G}_i^\top\mathbf{Z} \tag{3.10}$$

for each variable. This alternating procedure is continued until the differences between the values of the loss function (3.1) in the current and previous steps is less than ε.

The rank-restricted MCA algorithm is summarized as follows:

Step 1: Initial values are chosen for $\mathbf{A}_i$, $\mathbf{Z}$, and $\mathbf{A}_i$. Take arbitrary matrix $\mathbf{Q}_i$ which satisfies the constraints (3.3). Values for $\mathbf{Z}$ can be chosen randomly, which should satisfy the constraints. Then, $\mathbf{A}_i$ is given as $\mathbf{A} = n^{-1}\mathbf{Q}_i^\top \mathbf{G}_i^\top \mathbf{Z}$ for each variable.

Step 2: Update the quantification parameters by solving the orthogonal Procrustes problems.

Step 3: Update the MCA model parameters by solving the orthogonal Procrustes and regression problems.

Step 4: Finish if the differences between the values of the loss function (3.1) in the current and previous steps is less than ε; otherwise, return to *Step 2*.

The proposed algorithm monotonically decreases the loss function. Since the loss function is bounded below, it converges to a solution that is at least a local optimum (Young 1981).

To increase the chance of finding the global maximum, the algorithm should be run several times, with different initial values.

3.5 Numerical Example

We demonstrate MCA using the sleeping bag data (Prediger 1997) already given in Chap. 2. The original dataset contains numerical variables, and MCA cannot be applied to that type of dataset. Thus, we use three variables (Material, Quality Rate, and Price), and categorize Price to "Cheap," "not expensive," and "expensive." The dataset we modified is described in Table 3.2. For this demonstration, we report the $r = 2$ rank-restricted MCA solution. We set the dimension of quantification for Quality Rate and Price to be one, because they are considered as ordinal variables. In contrast, Material is considered to be nominal variable, and thus, the dimension of quantification is set to be $\min(K_1 - 1, 2) = 2$.

Tables 3.3, 3.4 and 3.5 report the quantified values of Material, Quality Rate and Price. Material is quantified without rank restriction due to being the nominal variable. The component scores and factor loadings are given in Tables 3.6 and 3.7, respectively. The MC result can be interpreted in the same way as an PCA result.

Table 3.2 Dataset (modified)

Bag	Price	Material	Quality rate
One kilo bag	Cheep	Liteloft	3
Sund	Cheep	Hollow ber	1
Kompakt basic	Cheep	MTI Loft	3
Finmark tour	Cheep	Hollow ber	1
Interlight lyx	Cheep	Thermolite	1
Kompakt	Not expensive	MTI Loft	2
Touch the cloud	Not expensive	Liteloft	2
Cat's meow	Not expensive	Polarguard	3
Igloo super	Not expensive	Terraloft	1
Donna	Not expensive	MTI Loft	2
Tyin	Not expensive	Ultraloft	2
Travellers dream	Not expensive	Goose-downs	3
Yeti light	Not expensive	Goose-downs	3
Climber	Not expensive	Duck-downs	2
Viking	Not expensive	Goose-downs	3
Eiger	Expensive	Goose-downs	2
Climber light	Not expensive	Goose-downs	3
Cobra	Expensive	Duck-downs	3
Cobra comfort	Expensive	Duck-downs	2
Fox re	Expensive	Goose-downs	3
Mont blanc	Expensive	Goose-downs	3

Table 3.3 Quantification matrix of Price

	DIM1
Cheep	1.747
Expensive	−0.913
Not expensive	−0.379

Table 3.4 Quantification matrix of material

	DIM1	DIM2
Duck-downs	0.673	0.548
Goose-downs	0.487	0.544
Hollow ber	−2.177	−0.560
Liteloft	0.245	−1.631
MTI Loft	0.388	−1.134
Polarguard	0.384	0.236
Terraloft	−1.608	2.799
Thermolite	−2.177	−0.560
Ultraloft	0.675	−0.139

Table 3.5 Quantification matrix of Quality rate

	DIM1
1	2.04
2	−0.68
3	−0.34

Table 3.6 Component scores of MCA

	Comp1	Comp2
One kilo bag	−0.116	−1.990
Sund	−2.143	−0.473
Kompakt basic	−0.091	−1.634
Finmark tour	−2.143	−0.473
Interlight Lyx	−2.143	−0.473
Kompakt	0.615	−0.871
Touch the cloud	0.591	−1.226
Cat's meow	0.378	0.218
Igloo super	−1.577	2.801
Donna	0.615	−0.871
Tyin	0.664	−0.160
Travellers dream	0.399	0.436
Yeti light	0.399	0.436
Climber	0.617	0.350
Viking	0.399	0.436
Eiger	0.683	0.574
Climber light	0.399	0.436
Cobra	0.615	0.632
Cobra comfort	0.758	0.560
Fox re	0.540	0.646
Mont blanc	0.540	0.646

Table 3.7 Component loadings of MCA

	Comp1	Comp2
Material (DIM1)	0.984	−0.035
Material (DIM2)	0.002	0.981
Quality rate	0.980	−0.183
Price	0.752	0.588

References

Adachi, K., Murakami, T.: Nonmetric Multivariate Analysis: MCA, NPCA, and PCA. Asakura Shoten, Tokyo (2011). (in Japanese)

Benzecri, J.P.: L'analyses des donnees: Tome (VoL) 1. La taxinomie: Tome. 2 La'analyses des correspondances, Dunod, Paris (1974)

Benzecri, J.P.: Correspondence Analysis Handbook. Marcel Dekker, New York (1992)

de Leeuw, J., van Rijckevorsel, J.L.A.: Beyond homogeneity analysis. In: van Rijckevorsel, J.L.A., de Leeuw, J. (eds.) Component and Correspondence Analysis: Dimension Reduction by Functional Approximation, pp. 55–80. Wiley, New York (1988)

Greenacre, M.J.: Theory and Application of Correspondence Analysis. Academic Press, London (1984)

Gifi, A.: Nonlinear Multivariate Analysis. Wiley, Chichester (1990)

Hayashi, C.: On the prediction of phenomena from qualitative data and the quantification of qualitative data from the mathematico-statistical point of view. Ann. Inst. Stat. Math. **3**, 69–98 (1952)

Jolliffe, L.T.: Principal Component Analysis, 2nd edn. Springer, New York (2002)

Murakami, T.: A Psychometrics Study on Principal Component Analysis of Categorical Data, Technical report (1999) (in Japanese)

Murakami, T., Kiers, H.A.L., ten Berge, J.M.F.: Non-metric principal component analysis for categorical variables with multiple quantifications, Unpublished manuscript (1999)

Nishisato, S.: Multidimensional Nonlinear Descriptive Analysis. Chapman and Hall, London (2006)

Prediger, S.: Symbolic objects in formal concept analysis, In: Mineau, G., Fall, A. (eds.) Proceedings of the 2nd International Symposium on Knowledge, Retrieaval, Use, and Storage for Efficiency (1997)

ten Berge, J.M.F.: Least Squares Optimazation in Multivariate Analysis. DSNO Press, Leiden (1993)

Tenenhaus, M., Young, Y.W.: An analysis and synthesis of multiple correspondence analysis, optimal scaling, dualscaling, homogeneity analysis and other methods for quantifying categorical multivariate data. Psychom. **50**, 91–119 (1985)

Young, F.W.: Quantitative analysis of qualitative data. Psychom. **46**, 357–388 (1981)

Part II
Applications and Related Topics

The second part of this book consists of four chapters, which examine applications of nonlinear principal component analysis (PCA). Four applications are introduced: variable selection for mixed measurement level data, sparse multiple correspondence analysis (MCA), reduced k-means clustering, and acceleration of alternating least squares (ALS) computation. The variable selection methods in PCA, which was originally developed for numerical data, can be applied to any measurement level data by using nonlinear PCA. Sparseness and k-means clustering for nonlinear data, which have been investigated in recent studies, are extensions obtained by the same matrix operations used in nonlinear PCA and MCA. Finally, an acceleration algorithm is proposed to reduce the computational cost of ALS iteration in nonlinear multivariate methods.

Chapter 4
Variable Selection in Nonlinear Principal Component Analysis

Abstract Chapter 2 shows that any measurement level multivariate data can be uniformly dealt with as numerical data in the context of principal component analysis (PCA) by using the alternating least squares with optimal scaling. This means that all variables in the data can be analyzed as numerical variables, and, therefore, we can solve the variable selection problem for mixed measurement level data using any existing variable selection method developed for numerical variables. In this chapter, we discuss variable selection in nonlinear PCA. We select a subset of variables that represents all variables as far as possible from mixed measurement level data using criteria in the modified PCA, which naturally includes a variable selection procedure.

Keywords Modified PCA · Stepwise selection · Cumulative proportion · RV-coefficient

4.1 Introduction

In many data analysis situations, categorical (nominal and ordinal) and numerical data are collected simultaneously. A typical approach for such mixed measurement level data is to analyze categorical data and numerical data separately by a method specified for each measurement level. For cases in which both data types are treated simultaneously, categorical variables are quantified by translating them to dummy variables before the numerical analysis, or categorical codes in each item are used as numerical variables in the analysis without any translation. However, these approaches are not suitable for cases in which we wish to use all of the collected variables to observe all data in lower dimensions and/or to select a subset of variables that represents all variables to the extent possible.

As shown in Chap. 2 describing the basic concept of principal component analysis (PCA) for mixed measurement level data, nonlinear PCA including optimal scaling quantifies all categorical variables in the data as homogeneous numerical variables. This means that we can handle all variables uniformly as numerical variables. In other words, we can apply ordinary multivariate methods developed for numerical data to the collected data.

© The Author(s) 2016
Y. Mori et al., *Nonlinear Principal Component Analysis and Its Applications*,
JSS Research Series in Statistics, DOI 10.1007/978-981-10-0159-8_4

Several studies have investigated variable selection in ordinary PCA for numerical variables. Jolliffe (1972, 1973) considered principal component (PC) loadings to find important variables. McCabe (1984) and Falguerolles and Jmel (1993) used a partial covariance matrix to select a subset of variables that maintains the information on all variables to the greatest extent possible. Robert and Escoufier (1976) and Bonifas et al. (1984) used the RV-coefficient, and Krzanowski (1987a, b) used Procrustes analysis to evaluate the similarity between the configuration of PCs computed based on selected variables and that based on all variables. Tanaka and Mori (1997) discussed a method called modified PCA (M.PCA) that can be used to compute PCs using only a select subset of variables that represents all of the variables, including those not selected. Since M.PCA includes variable selection procedures in the analysis, its criteria can be used directly to detect a reasonable subset of variables (e.g., Mori et al. 2006). Fueda et al. (2009) estimated PCs based on a subset of variables and selected a reasonable subset of variables using the estimation technique. Furthermore, PCA with sparse loadings, such as SCoTLASS by Zou et. al. (2004) and sparse PCA by Jolliffe et al. (2003), can be used to identify sets of important variables. Group sparse PCA and sparse multiple correspondence analysis by Bernard et al. (2012) also select groups of numerical variables and categorical variables. (Note that the approach to categorical variables of Bernard et al. (2012) differs from our approach, as described in Chap. 5.)

Among them, we focus on variable selection using the criteria in M.PCA (Tanaka and Mori 1997) to select a subset of variables from mixed measurement level data because M.PCA naturally includes variable selection in the PCA process and is easily extended to nonlinear M.PCA. The concept and procedures for selecting a subset of numerical variables using the criteria in M.PCA are summarized in Mori et al. (2006), while the basic concept was introduced in Tanaka and Mori (1997). Moreover, the selection procedures and computational environment were proposed in Mori et al. (1998) and Mori et al. (2000), respectively. The initial concept for selecting a subset of categorical variables using the criteria in M.PCA, which is the *Type 1* selection described in Sect. 4.3.2, was proposed in Mori et al. (1997). Computations were performed in Kuroda et al. (2011) as an application of accelerated computation for nonlinear PCA (acceleration is introduced in Chap. 7). Here, we formulate a variable selection by applying the criteria in M.PCA to select a subset of variables from mixed measurement level data.

4.2 Modified PCA for Mixed Measurement Level Data

4.2.1 An Overview of Modified PCA

M.PCA (Tanaka and Mori 1997) is intended to derive PCs that are computed using only a selected subset but represent all of the variables, including those not selected. Those PCs provide a multidimensional rating scale which has high validity and is easy

to apply practically, although it is obtained from small numbers of variables. In order to find such PCs we can borrow the concepts of Rao (1964)'s PCA of instrumental variables and Robert and Escoufier (1976)'s RV-coefficient-based approach.

Suppose we obtain an $n \times p$ data matrix $\mathbf{Y}$ that consists of numerical variables. Let $\mathbf{Y}$ be decomposed into an $n \times q$ submatrix $\mathbf{Y}_1$ and an $n \times (p - q)$ submatrix $\mathbf{Y}_2$ $(1 \le q \le p)$. We denote the covariance matrix of $\mathbf{Y} = (\mathbf{Y}_1, \mathbf{Y}_2)$ as $\mathbf{S} = \begin{pmatrix} \mathbf{S}_{11} & \mathbf{S}_{12} \\ \mathbf{S}_{21} & \mathbf{S}_{22} \end{pmatrix}$, where the subscript i of $\mathbf{S}$ corresponds to $\mathbf{Y}_i$. $\mathbf{Y}$ is represented as accurately as possible by r PCs, where r is the number of PCs and the PCs are linear combinations of a submatrix $\mathbf{Y}_1$, i.e., $\mathbf{Z} = \mathbf{Y}_1 \mathbf{A}$ $(1 \le r \le q)$. In order to derive $\mathbf{A} = (\mathbf{a}_1, \ldots, \mathbf{a}_r)$, the following Criterion 1 based on Rao (1964) and Criterion 2 based on Robert and Escoufier (1976) can be used:

(Criterion 1) The prediction efficiency for $\mathbf{Y}$ is maximized using a linear predictor in terms of $\mathbf{Z}$.
(Criterion 2) The RV-coefficient between $\mathbf{Y}$ and $\mathbf{Z}$ is maximized. The RV-coefficient is computed as

$$RV(\mathbf{Y}, \mathbf{Z}) = \mathrm{tr}(\tilde{\mathbf{Y}}\tilde{\mathbf{Y}}^\top \tilde{\mathbf{Z}}\tilde{\mathbf{Z}}^\top)/\{\mathrm{tr}(\tilde{\mathbf{Y}}\tilde{\mathbf{Y}}^\top)\mathrm{tr}(\tilde{\mathbf{Z}}\tilde{\mathbf{Z}}^\top)\}^{1/2},$$

where $\tilde{\mathbf{Y}}$ and $\tilde{\mathbf{Z}}$ are centered matrices of $\mathbf{Y}$ and $\mathbf{Z}$, respectively.

The maximization criteria for the above (Criterion 1) and (Criterion 2) are given by the proportion P

$$P = \sum_{j=1}^{r} \lambda_j / \mathrm{tr}(\mathbf{S}), \tag{4.1}$$

and the RV-coefficient

$$RV = \left\{ \sum_{j=1}^{r} \lambda_j^2 / \mathrm{tr}(\mathbf{S}^2) \right\}^{1/2}, \tag{4.2}$$

s respectively, where λ_j is the j-th eigenvalue with the order of magnitude of the eigenvalue problem (EVP)

$$[(\mathbf{S}_{11}^2 + \mathbf{S}_{12}\mathbf{S}_{21}) - \lambda \mathbf{S}_{11}]\mathbf{a} = 0. \tag{4.3}$$

When the number of variables in $\mathbf{Y}_1$ is q, $\mathbf{Y}_1$ should be assigned by a subset of q variables ($\mathbf{Y}_2$ by a subset of $p - q$ remaining variables) that provides the largest value of P by Eq. (4.1) for (Criterion 1) or the largest value of RV by Eq. (4.2) for (Criterion 2), and the solution is obtained as a matrix $\mathbf{A}$, the columns of which consist of the eigenvectors associated with the largest r eigenvalues of EVP (4.3).

Obviously, these criteria can be used to select a reasonable subset of size q; in other words, "variable selection using the criteria in M.PCA" is to find a subset of size q by searching for one that has the largest value of the above criterion P or RV among all possible subsets of size q.

4.2.2   Modified PCA for Mixed Measurement Level Data

Since M.PCA is a method for numerical variables, categorical variables in the data should be quantified in an appropriate manner. This means that we implement M.PCA for categorical variables (nonlinear M.PCA) as in Sect. 2.1. Considering PRINCI-PALS in Sect. 2.3.3.1, the implementation is easy because nonlinear M.PCA can be formulated by replacing only the eigen-decomposition in *Model estimation step* of PRINCIPALS in Sect. 2.3.3.1 by the EVP (4.3) to get the model parameters $\mathbf{A}$ and $\mathbf{Z}$ for M.PCA (Mori et al. 1997).

Here, we rewrite the ALS algorithm of PRINCIPALS as follows: for given initial data $\mathbf{Y}^{*(0)} = (\mathbf{Y}_1^{*(0)}, \ \mathbf{Y}_2^{*(0)})$ from original data $\mathbf{Y}$, the following two steps are iterated until convergence (i.e., $\theta^* = \mathrm{tr}(\mathbf{Y}^* - \hat{\mathbf{Y}})^\top(\mathbf{Y}^* - \hat{\mathbf{Y}}) = \mathrm{tr}(\mathbf{Y}^* - \mathbf{Z}\mathbf{A}^\top)^\top(\mathbf{Y}^* - \mathbf{Z}\mathbf{A}^\top)$ is minimized):

- *Model estimation step*: From $\mathbf{Y}^{*(t)} = (\mathbf{Y}_1^{*(t)}, \ \mathbf{Y}_2^{*(t)})$, obtain $\mathbf{A}^{(t)}$ by solving an eigenvalue problem

$$[(\mathbf{S}_{11}^2 + \mathbf{S}_{12}\mathbf{S}_{21}) - \lambda\mathbf{S}_{11}]\mathbf{a} = 0.$$

Compute $\mathbf{Z}^{(t)}$ from $\mathbf{Z}^{(t)} = \mathbf{Y}_1^{*(t)}\mathbf{A}^{(t)}$.
- *Optimal scaling step*: Calculate $\hat{\mathbf{Y}}^{(t+1)} = \mathbf{Z}^{(t)}\mathbf{A}^{(t)\top}$. Find $\mathbf{Y}^{*(t+1)}$ such that

$$\mathbf{Y}^{*(t+1)} = \arg\min_{\mathbf{Y}^{*(t)}} \mathrm{tr}(\mathbf{Y}^{*(t)} - \hat{\mathbf{Y}}^{(t+1)})^\top(\mathbf{Y}^{*(t)} - \hat{\mathbf{Y}}^{(t+1)})$$

for fixed $\hat{\mathbf{Y}}^{(t+1)}$ under measurement restrictions on each of the variables. Since $\mathbf{Y}^{*(t+1)}$ is obtained by separately estimating $\mathbf{Y}_j^*$ for each j ($j = 1, \ldots, p$), scale $\mathbf{Y}^{*(t+1)}$ by columnwise centering and normalizing. Re-compute $\mathbf{Y}_j^{(t+1)}$ by an additional transformation to keep the monotonicity restriction for ordinal variables and skip this computation for numerical variables.

$\mathbf{Y}^* = (\mathbf{Y}_1^*, \ \mathbf{Y}_2^*)$ obtained after convergence is an optimally scaled (quantified) matrix of $\mathbf{Y}$.

4.3   Variable Selection in Nonlinear Modified PCA

4.3.1   Four Selection Procedures

When the size of $\mathbf{Y}_1$ is q, the best subset of q variables selected by using a criterion in M.PCA is one that provides the largest criterion value P or RV among all possible ${}_pC_q$ combinations of variables. Although finding such a subset among all possible subsets is the best selection procedure, this procedure is usually impractical due to the high computational cost of computing criterion values for all possible subsets.

Therefore, as practical strategies, we can use one of four selection procedures (see, e.g., Mori et al. 2006):

a. Backward elimination
b. Backward–forward stepwise selection
c. Forward selection
d. Forward–backward stepwise selection

in which only one variable is removed or added sequentially. These procedures allow automatic selection of any number of variables.

Each procedure consists of two stages. The one below is the detailed flow reproduced from Mori et al. (2006), where V is the criterion value P or RV obtained by assigning q variables to $\mathbf{Y}_1$:

a. Backward elimination

Stage A.   Initial fixed-variable stage

A-1   Assign q variables to subset $\mathbf{Y}_1$, usually $q := p$.
A-2   Solve the EVP (4.3).
A-3   Look carefully at the eigenvalues, determine the number r of PCs to be used.
A-4   Specify kernel variables that should always be involved in $\mathbf{Y}_1$, if necessary. The number of kernel variables is less than q.

Stage B.   Variable selection stage (Backward)

B-1   Remove one variables from among q variables in $\mathbf{Y}_1$, make a temporary subset of size $q - 1$, and compute V based on the subset. Repeat this for each variable in $\mathbf{Y}_1$, then obtain q Vs. Find the best subset of size $q - 1$ that provides the largest V among q Vs and remove the corresponding variable from the present $\mathbf{Y}_1$. Put $q := q - 1$.
B-2   If the V or q is larger (or smaller) than the preassigned values, go to B-1. Otherwise stop.

b. Backward–forward stepwise selection

Stage A.   Initial fixed-variable stage

A-1 $\sim$ 4   Same as A-1 to 4 in Backward elimination.

Stage B.   Variable selection stage (Backward–forward)

B-1   Put $i := 1$.
B-2   Remove one variable from among q variables in $\mathbf{Y}_1$, make a temporary subset of size $q - 1$, and compute V based on the subset. Repeat this for each variable in $\mathbf{Y}_1$, then obtain q Vs. Find the best subset of size $q - 1$ that provides the largest V (denoted by V_i) among q Vs and remove the corresponding variable from the present $\mathbf{Y}_1$. Set $q := q - 1$.

B-3   If the V or q is larger (or smaller) than preassigned values, go to B-4. Otherwise stop.

B-4   Remove one variable from among q variables in $\mathbf{Y}_1$, make a temporary subset of size $q - 1$, and compute V based on the subset. Repeat this for each variable in $\mathbf{Y}_1$, then, obtain q Vs. Find the best subset of size $q - 1$ that provides the largest V (denoted by V_{i+1}) among q Vs and remove the corresponding variable from the present $\mathbf{Y}_1$. Set $q := q - 1$.

B-5   Add one variable from among $p - q$ variables in $\mathbf{Y}_2$ to $\mathbf{Y}_1$, make a temporary subset of size $q + 1$, and compute V based on the subset. Repeat this for each variable, except for the variable removed from $\mathbf{Y}_1$ and moved to $\mathbf{Y}_2$ in B-4, then obtain $p - q - 1$ Vs. Find the best subset of size $q + 1$ that provides the largest V (denoted by V_{temp}) among $p - q - 1$ Vs.

B-6   If $V_i < V_{temp}$, add the variable found in B-5 to $\mathbf{Y}_1$, set $V_i := V_{temp}$, $q := q + 1$ and $i := i - 1$, and go to B-5. Otherwise set $i := i + 1$ and go to B-3.

c. Forward selection

Stage A.   Initial fixed-variable stage

A-1 $\sim$ 3   Same as A-1 to 3 in Backward elimination.

A-4   Redefine q as the number of kernel variables (here, $q \geq r$). If you have kernel variables, assign them to $\mathbf{Y}_1$. If not, put $q := r$, find the best subset of q variables that provides the largest V among all possible subsets of size q and assign it to $\mathbf{Y}_1$.

Stage B.   Variable selection stage (Forward)
Basically, the opposite of Stage B in Backward elimination.

d. Forward–backward stepwise selection

Stage A.   Initial fixed-variable stage

A-1 to 4   Same as A-1 to 4 in Forward selection.

Stage B.   Variable selection stage (Forward–backward)
Basically, the opposites of Stage B in Backward–forward stepwise selection.

The criteria based on the subsets selected by the above procedures differ only slightly from those based on the best subset among all possible ones. The stepwise-type selections (b and d) can select better subsets than the single-type selections (a and c) and the forward-type selections (c and d) tend to select better subsets than the backward-type selections (a and b).

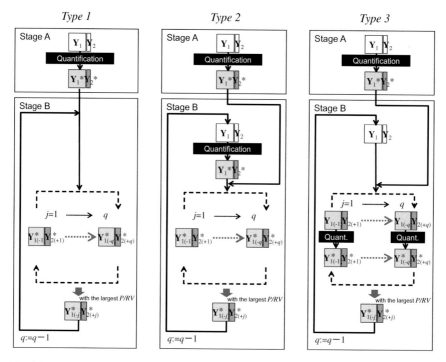

Fig. 4.1   Three possible *types* of selection (this flow illustrates the case of backward-type selection)

4.3.2   Three Possible Types of Selection

The selection flows in Sect. 4.3.1 are for data $\mathbf{y}$ containing only numerical variables, i.e., all variables in $\mathbf{Y}$ are numerical. For data $\mathbf{Y}$ containing categorical variables, we have to implement quantification (optimal scaling with ALS; here, nonlinear M.PCA) in the flow. There are three *types* of selection (*Type 1*, *Type 2* and *Type 3* shown below) according to where the quantification of mixed measurement level data $\mathbf{Y}$ is implemented in the flow (Fig. 4.1).

For all selection *types* in common, q original variables are assigned to $\mathbf{Y}_1$ at A-1 and nonlinear M.PCA is applied to $(\mathbf{Y}_1, \mathbf{Y}_2)$ at A-2. Then quantified $(\mathbf{Y}_1^*, \mathbf{Y}_2^*)$ is obtained and the remaining A-3 and A-4 in Stage A are done for this quantified data.

In *Type 1*, no more quantification is carried out in Stage B, i.e., all processes in Stage B are performed on $\mathbf{Y}^*$ which has been quantified in Stage A. The quantification is performed only once through the Stage A and B. (See the left flow in Fig. 4.1.)

In *Type 2*, the quantification is carried out every time after the best subset of size q is found at B-2 of the single-type selections, B-2, B-4, and B-5 of the stepwise-type selections. That is, the quantified $(\mathbf{Y}_1^*, \mathbf{Y}_2^*)$ based on the best subset of size q found in the previous selection is used to find the best subset of size $q-1$ or $q+1$ in the next selection. (See the middle flow in Fig. 4.1.)

In *Type 3*, the quantification is carried out for every temporary $(\mathbf{Y}_1, \mathbf{Y}_2)$ in Stage B, i.e., nonlinear M.PCA is executed whenever temporary $(\mathbf{Y}_1, \mathbf{Y}_2)$ is given to compute its criterion value V in Stage B. (See the right flow in Fig. 4.1.)

A reasonable subset of size q is given as $\mathbf{Y}_1$ corresponding to the best subset $\mathbf{Y}_1^*$ which is finally found at q when the selection procedure is terminated.

4.4   Numerical Examples

We use a data set of alate adelges (winged aphids) as a demonstration data, which was originally analyzed by Jeffers (1967) using ordinary PCA. The data set consists of 40 individuals and 19 variables. The variables are all numerical, which measure 19 physical sizes. Eigenvalues and their cumulative proportions of the data are 13.8379 (72.83%), 2.3635 (85.27%), 0.7480 (89.21%), ..., then two PCs ($r = 2$) were used in previous studies and are used here.

Here, we modify this data set to an artificial data set with mixed measurement level variables as follows: we translate V18 (anal fold: exits or not) to nominal and {V6, ..., V10, V12, ..., V14} to ordinal in five levels. Then, the obtained artificial data set consists of 1 nominal, 8 ordinal, and 10 numerical variables (Table 4.1).

Since Jeffers (1967) found four clusters by observing the plot of PCs obtained by ordinary PCA based on the correlation matrix of whole variables, we choose the RV-coefficient as a selection criterion to detect a subset providing the close configuration of PCs to the original configuration.

Figures 4.2 and 4.3 indicate biplots of PC scores and loadings obtained by applying M.PCA to the original and artificial data, respectively. Figure 4.2 is the same plot as the PC scores plot obtained by ordinary PCA because M.PCA provides the same results as ordinary PCA when all variables $(\mathbf{Y} = \mathbf{Y}_1)$ are used. It is not our purpose here to confirm whether similar results are obtained from two data sets but these plots indicate that the same four clusters are observed, although the loadings are slightly different because the features of some variables are changed from the original. This indicates that nonlinear PCA (M.PCA) reproduces the original configuration well.

We apply *Type 3* of Forward–backward stepwise selection to the artificial deta set. The procedure identifies a reasonable subset for each q ($q = p \ldots r$) in turn.

Table 4.2 shows the selection results of $(\mathbf{Y}_1, \mathbf{Y}_2)$ and RV-coefficients for every q. The change of the RV-coefficients is shown in Fig. 4.4. (although the selection was executed from $q = r$ to p because of forward-type selection, the results in Table 4.2 and the change of criterion value in Fig. 4.4 are indicated from $q = p$ to r.)

The results illustrate that the RV-coefficients change slightly when the number of variables exceeds five. For example, when we delete five variables, the difference between RV-coefficient of the original 19 variables (0.9910) and that of the selected 14 variables (0.9894) is very small (0.0016). When we delete 13 variables, the difference is 0.0112. In particular, the sequential difference is less than 0.001 until the number of variables is nine.

Table 4.1 Artificial data (modified from Jeffers (1967)'s alate adelges data)

No	V1 (length)	V2 (width)	V3 (fwing)	V4 (hwing)	V5 (numspi)	V6 (antseg1)	V7 (antseg2)	V8 (ntseg3)	V9 (antseg4)	V10 (antseg5)	V11 (numaspi)	V12 (tarsus)	V13 (tibia)	V14 (femur)	V15 (rostrum)	V16 (ovipos)	V17 (ovispi)	V18 (anal)	V19 (numhooks)
1	21.2	11	7.5	4.8	5	4	5	5	5	5	3	5	5	5	7	4	8	0	3
2	20.2	10	7.5	5	5	5	5	5	5	4	5	5	5	5	7.6	4.2	8	0	3
3	20.2	10	7	4.6	5	3	5	5	4	5	1	5	5	5	7	4	6	0	3
4	22.5	8.8	7.4	4.7	5	5	5	5	4	4	4	5	5	5	6.8	4.1	6	0	3
5	20.6	11	8	4.8	5	5	5	5	4	4	5	5	5	5	6.7	4	6	0	3
6	19.1	9.2	7	4.5	5	3	5	5	5	4	4	5	5	5	5.7	3.8	8	0	3.5
7	20.8	11.4	7.7	4.9	5	5	5	5	5	4	4	5	5	5	6.6	4	8	0	3
8	15.5	8.2	6.3	4.9	5	4	5	5	3	4	3	4	4	4	6.7	3.5	6	0	3.5
9	16.7	8.8	6.4	4.5	5	4	5	5	4	4	3	4	4	3	6.1	3.7	8	0	3
10	19.7	9.9	8.2	4.7	5	4	5	5	5	4	0	5	5	5	6	3.8	8	0	3
11	10.6	5.2	3.9	2.3	4	1	1	2	2	1	6	1	1	1	4.5	2.7	4	1	2
12	9.2	4.5	3.7	2.2	4	1	1	2	1	1	5	1	1	1	4.1	2.4	4	1	2
13	9.6	4.5	3.6	2.3	4	1	1	2	1	1	4	1	1	1	4	2.3	4	1	2
14	8.5	4	3.8	2.2	4	1	1	2	2	1	5	1	1	1	4.4	2.3	4	1	2
15	11	4.7	4.2	2.3	4	1	1	2	2	1	4	1	1	1	4.5	2.6	4	1	2
16	18.1	8.2	5.9	3.5	5	3	5	2	4	3	4	3	4	4	6	4.5	9	1	2
17	17.6	8.3	6	3.8	5	4	5	2	3	3	3	3	3	3	5.7	4.3	10	1	2
18	19.2	6.6	6.2	3.4	5	4	4	3	3	3	4	3	3	3	5.3	3.7	10	1	2
19	15.4	7.6	7.1	3.4	5	4	5	4	4	3	4	3	3	3	6	4.2	8	1	3
20	15.1	7.3	6.2	3.8	5	4	4	3	3	2	4	4	3	3	6.4	4.3	10	1	2.5
21	16.1	7.9	5.8	3.7	5	4	5	3	4	3	5	4	3	3	6	4.5	10	1	2
22	19.1	8.8	6.4	3.9	5	4	5	3	3	3	4	4	4	4	6.5	4.5	10	1	2.5
23	15.3	6.4	5.3	3.3	5	2	5	2	3	2	4	3	3	3	5.4	4	10	1	2
24	14.8	8.1	6.2	3.7	5	4	5	3	3	4	4	4	3	3	6	4.1	10	1	2
25	16.2	7.7	6.9	3.7	5	4	4	3	3	2	5	3	3	3	5.7	4.2	10	1	2.5
26	13.4	6.9	5.7	3.4	5	4	4	5	2	2	4	4	2	2	6	3.9	10	1	2
27	12.9	5.8	4.8	2.6	5	2	3	2	2	2	5	2	2	2	5.5	3.6	9	1	3
28	12	6.5	5.3	3.2	5	3	5	3	4	4	5	3	3	3	5.1	4.3	8	1	2
29	14.1	7	5.5	3.6	5	4	5	3	4	4	5	4	3	3	5.8	4.1	10	1	2

(continued)

Table 4.1 (continued)

No	V1 (length)	V2 (width)	V3 (fwing)	V4 (hwing)	V5 (numspi)	V6 (antseg1)	V7 (antseg2)	V8 (ntseg3)	V9 (antseg4)	V10 (antseg5)	V11 (numaspi)	V12 (tarsus)	V13 (tibia)	V14 (femur)	V15 (rostrum)	V16 (ovipos)	V17 (ovispi)	V18 (anal)	V19 (numhooks)
30	16.7	7.2	5.7	3.5	5	3	5	4	3	3	5	3	3	3	6	4	10	1	2.5
31	14.1	5.4	5	3	5	2	3	2	4	2	5	1	2	2	5.3	3.6	8	1	2
32	10	6	4.2	2.5	5	2	2	1	2	3	6	2	1	1	4.8	3.4	8	1	2
33	11.4	4.5	4.4	2.7	5	3	3	2	1	2	5	1	1	1	4.7	3.7	8	1	2
34	12.5	5.5	4.7	2.3	5	3	2	2	3	2	4	2	1	1	5.1	3.7	8	0	2
35	13	5.3	4.7	2.3	5	2	2	2	1	1	4	1	1	2	5	3.6	8	1	2
36	12.4	5.2	4.4	2.6	5	2	2	1	3	2	5	1	1	1	5	3.2	6	1	2
37	12	5.4	4.9	3	5	2	3	2	1	1	5	1	1	1	4.2	3.7	6	1	2
38	10.7	5.6	4.5	2.8	5	3	2	2	3	2	4	1	1	1	5	3.5	8	1	2
39	11.7	5.5	4.3	2.6	5	2	3	2	1	2	5	1	1	1	4.6	3.4	8	1	2
40	12.8	5.7	4.8	2.8	5	2	2	1	1	1	5	1	1	1	5	3.1	8	1	2

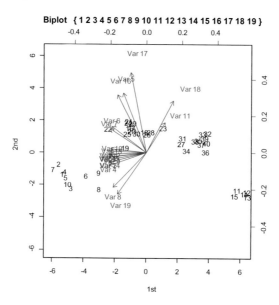

Fig. 4.2 Biplots of original data

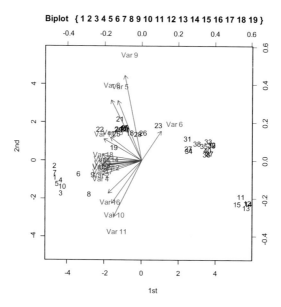

Fig. 4.3 Biplots of artificial data

Table 4.2 Selection results (Artificial data, Forward–backward selection, $r = 2$, RV-coefficient)

q	$\mathbf{Y}_i$	$\mathbf{Y}_2$																	RV-coef.
19	1	2	3	4	5	6	7	8	9	10	11	12	13	14	15	16	17	18	19	0.9910055
18	1	2	3	4	5	6	7	8	9	10	11	12	13	14	15	16	17	18	|19	0.9909302
17	1	2	3	4	5	6	7	8	9	10	11	12	13	14	15	16	17	|18	19	0.9905681
16	1	3	4	5	6	7	8	9	10	11	12	13	14	15	16	17	|2	18	19	0.9903543
15	1	3	4	6	6	7	8	9	10	11	12	13	15	16	17	|2	14	18	19	0.9899835
14	1	3	4	5	6	7	9	10	11	12	13	15	16	17	|2	8	14	18	19	0.9894428
13	1	3	5	7	8	9	10	11	12	13	15	16	17	|2	8	11	14	18	19	0.9885673
12	3	4	5	7	8	9	11	13	13	15	16	17	|1	2	8	12	14	18	19	0.9880060
11	3	4	5	7	8	9	11	13	15	16	17	|1	2	6	9	12	14	18	19	0.9874418
10	3	4	7	8	9	11	13	15	15	17	|1	2	6	10	11	14	16	18	19	0.9869801
9	3	4	8	9	11	13	15	17	17	|1	2	5	6	10	12	14	16	18	19	0.9862902
8	3	7	8	9	11	13	18	|1	|1	2	4	5	6	10	12	14	16	18	19	0.9841846
7	3	7	9	11	13	18	|1	2	2	4	5	6	10	12	14	15	16	17	19	0.9823263
6	3	7	11	13	18	|1	2	4	4	5	6	8	10	12	14	15	16	17	19	0.9797715
5	3	9	8	17	|1	2	5	6	5	6	7	8	10	12	14	15	16	17	19	0.9704375
4	3	4	8	|1	2	2	5	6	7	9	10	11	13	13	14	15	16	18	19	0.9584721
3	3	4	8	|1	2	5	6	7	9	10	11	12	13	14	15	16	17	18	19	0.9465730
2	4	8	|1	2	3	5	6	7	9	10	11	12	13	14	15	16	17	18	19	0.9206419

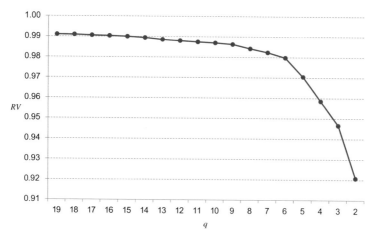

Fig. 4.4 Change of RV-coefficients

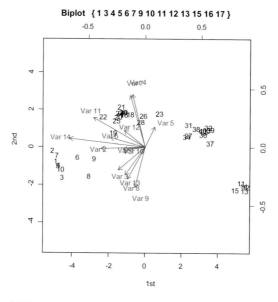

Fig. 4.5 Biplots of PC scores and loadings based on selected 14 variables

Here, we draw biplots of PC scores and loadings based on the selected 14 variables {V1, V3, V4, V5, V6, V7, V9, V10, V11, V12, V13, V15, V16, V17} (Fig. 4.5) and 6 variables {V3, V7, V9, V11, V13, V18} (Fig. 4.6). Their RV-coefficients are 0.9894 and 0.9798, respectively. Comparing these plots and RV-coefficients with those of 19 original variables (Fig. 4.3, $RV = 0.9910$), it is observed that the selected variables keep the configuration of PC scores based on the original 19 variables even though the number of variables is one third of the original number of variables. This means that

Fig. 4.6 Biplots of PC
scores and loadings based on
selected 6 variables

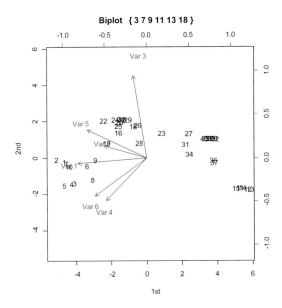

the proposed selection can select variables that provide almost the same information
as the original PC scores in the context of nonlinear M.PCA.

References

Bernard, A., Guinot, C., Saporta, G.: Sparse principal component analysis for multiblock data and
    its extension to sparse multiple correspondence analysis. Compstat **2012**, 99–106 (2012)
Bonifas, I., Escoufier, Y., Gonzalez, P.L. et Sabatier, R.: Choix de variables en analyse en com-
    posantes principales. Rev. Statist. Appl. **23**, 5–15 (1984)
Falguerolles, A., De et Jmel, S.: Un critere de choix de variables en analyse en composantes
    principales fonde sur des modeles graphiques gaussiens particuliers. Rev. Canadienne Statist.
    21(3), 239–256 (1993)
Fueda, K., Iizuka, M., Mori, Y.: Variable selection in multivariate methods using global score
    estimation. Comput. Stat. **24**, 127–144 (2009)
Jeffers, J.N.R.: Two case studies in the application of principal component analysis. Appl. Stat. **16**,
    225–236 (1967)
Jolliffe, I.T.: Discarding variables in a principal component analysis I - Artificial data. Appl. Stat.
    21, 160–173 (1972)
Jolliffe, I.T.: Discarding variables in a principal component analysis II - Real data -. Appl. Stat. **22**,
    21–31 (1973)
Jolliffe, I.T., Trendafilov, N.T., Uddin, M.: A modified principal component technique based on the
    Lasso. J. Comput. Graph. Stat. **12**, 531–547 (2003)
Kuroda, M., Iizuka, M., Mori, Y., Sakakihara, M.: Principal components based on a subset of
    qualitative variables and its accelerated computational algorithm. Invited Paper Session 42. The
    58th World Statistics Congress of the International Statistical Institute (ISI2011). Dublin, Ireland
    (2011)

Krzanowski, W.J.: Selection of variables to preserve multivariate data structure, using principal components. Appl. Statist. **36**, 22–33 (1987a)

Krzanowski, W.J.: Cross-validation in principal component analysis. Biometrics **43**, 575–584 (1987b)

McCabe, G.P.: Principal variables. Technometrics **26**, 137–44 (1984)

Mori, Y., Tanaka, T., Tarumi, T.: Principal component analysis based on a subset of qualitative variables. In: C. Hayashi et al. (eds.). Proceedings of IFCS-96: Data Science, Classification and Related Methods pp. 547–554. Springer (1997)

Mori, Y., Tarumi, T., Tanaka, Y.: Principal Component analysis based on a subset of variables - Numerical investigation on variable selection procedures. Bull. Comput. Stat. Jpn. **11**(1), 1–12 (1998) (in Japanese)

Mori, Y., Iizuka, M. Tarumi, T., Tanaka, Y.: Statistical software "VASPCA" for variable selection in principal component analysis. In: Jansen, W., Bethlehem, J.G. (eds.). Proceedings in Computational Statistics COMPSTAT2000 (Short Communications), pp. 73–74 (2000)

Mori, Y., Iizuka, M., Tanaka, Y., Tarumi, T.: Variable selection in principal component analysis, In: Härdle, W., Mori, Y., Vieu, P. (eds.) Statistical Methods for Biostatistics and Related Fields, pp. 265–283. Springer, Berlin (2006)

Rao, C.R.: The use and interpretation of principal component analysis in applied research. Sankhya **A26**, 329–358 (1964)

Robert, P., Escoufier, Y.: A unifying tool for linear multivariate statistical methods: the RV-coefficient. Appl. Statist. **25**, 257–65 (1976)

Tanaka, Y., Mori, Y.: Principal component analysis based on a subset of variables: variable selection and sensitivity analysis. Am. J. Math. Manag. Sci. **17**(1&2), 61–89 (1997)

Zou, H., Hastie, T., Tibshirani, R.: Sparse principal component analysis. J. Comput. Graph. Stat. **15g**, 265–286 (2004)

Chapter 5
Sparse Multiple Correspondence Analysis

Abstract In multiple correspondence analysis (MCA), an estimated solution can be transformed into a simple structure in order to simplify the interpretation. The rotation technique is widely used for this purpose. However, an alternative approach, called sparse MCA, has also been proposed. One of the advantages of sparse MCA is that, in contrast to unrotated or rotated ordinary MCA loadings, some loadings in sparse MCA can be exactly zero. A real data example demonstrates that sparse MCA can provide simple solutions.

Keywords Simple structure · Rotation · Penalty-free approach · Loading matrix

5.1 Introduction

In principal component analysis (PCA), the estimated solutions can be transformed without changing the fitness of the loss function

$$\|\mathbf{Y} - \mathbf{Z}\mathbf{A}^\top\|^2 = \|\mathbf{Y} - \mathbf{Z}\mathbf{S}^\top \mathbf{S}^{\top^{-1}} \mathbf{A}^\top\|^2, \tag{5.1}$$

where $\mathbf{Y}$ is an observed data matrix (n-observations $\times$ p-variables), $\mathbf{Z}$ is a component score matrix (n-observations $\times$ r-components), $\mathbf{A}$ is a loading matrix (p-variables $\times$ r-components), and $\mathbf{S}$ is an arbitrary nonsingular matrix (p-variables $\times$ p-variables). This non-uniqueness is referred to as rotational freedom or rotational indeterminacy. Generally, using this property, the estimated loading matrix is transformed into a desirable simple structure in order to simplify the interpretation, which is referred to as rotation (e.g., Browne 2001). Thurstone (1947) provided five rules for a simple structure:

1. Each row of the factor matrix should have at least one zero;
2. If there are r common factors, each column of the factor matrix should have at least r zeros;
3. For every pair of columns of the factor matrix, there should be several variables, the entries of which vanish in one column but not in the other;

© The Author(s) 2016
Y. Mori et al., *Nonlinear Principal Component Analysis and Its Applications*,
JSS Research Series in Statistics, DOI 10.1007/978-981-10-0159-8_5

4. For every pair of columns of the factor matrix, a large proportion of the variables should have vanishing entries in both columns when there are four or more factors;
5. For every pair of columns of the factor matrix, there should be only a small number of variables with non-vanishing entries in both columns.

In other words, the simple structure should have a large number of zero elements, making the estimated solution easy to interpret by focusing only on the variables with nonzero elements. If no identification condition is introduced to $\mathbf{S}$, an arbitrary nonsingular transformation can be applied to the solution. In order to reduce this non-uniqueness, the constraints $n^{-1}\mathbf{S}^{\top}\mathbf{S} = \mathbf{I}_r$ (orthogonal rotation) or $n^{-1}diag(\mathbf{S}^{\top}\mathbf{S}^{\top^{-1}}) = \mathbf{I}_r$ (oblique rotation) are usually imposed on $\mathbf{S}$. In the psychometrics literature, a number of rotation methods have been proposed. In these studies, most of the objective functions are mathematically defined by Thurstone's concept and are optimized for exploring the matrices $\mathbf{S}$ that can transform the loading matrix $\mathbf{A}$ into a simple structure (Browne 2001). The rotation technique is a commonly used method for obtaining a simple structure, but rotated solutions produce numerous small non-zero loadings. The common practice in interpreting the rotated loadings is to set all loadings having magnitudes lower than some threshold to zero. Trendafilov and Adachi (2015) pointed out that the original concept of Thurstone's simple structure requires numerous exactly zero entries and cannot be obtained by rotation methods. Recently, in the PCA literature, an alternative method called sparse PCA has been proposed for overcoming the drawback of rotation methods mentioned above (e.g., Trendafilov 2014; Zou et al. 2006). In contrast to unrotated or rotated ordinary PCA loadings, some loadings in sparse PCA can be exactly zero. Note that although the true simple structure can be obtained by sparse PCA, in contrast to the rotation, the fitness of the loss function is changed. For obtaining the true Thurstone's simple structure in the MCA solution, we proposed a sparse MCA method that follows the sparse PCA.

We introduce the rotation method in MCA in more detail and then discuss the sparse MCA method. A real data example is provided in the last section.

5.2 Rotation

MCA also has the rotational freedom property, and the rotation of MCA solutions has been dealt with in some studies (Adachi 2004; Kiers 1991; van de Velden and Kiers 2003). However, there are some differences between the rotation in PCA and MCA: the relation between the loadings and component scores are known for PCA, $\mathbf{A} = n^{-1}\mathbf{Z}^{\top}\mathbf{Y}$, but this does not generally hold in MCA (Kiers 1991). Thus, the discrimination measure matrix, which is defined as $\mathbf{W} = \sum \mathbf{W}_i\mathbf{D}_i\mathbf{W}_i$, is rotated towards the simple structure. However, as mentioned in Sect. 5.1, the loading matrix can be defined in the MCA formulation proposed by Murakami et al. (1999) in the same manner as PCA. The quantification matrix can also be rotated as well as the loading and component score matrices in MCA:

$$\sum_{i=1}^{p} \|\mathbf{G}_i \mathbf{Q}_i - \mathbf{Z}\mathbf{A}_i^\top\|^2 = \sum_{i=1}^{p} \|\mathbf{G}_i \mathbf{Q}_i \mathbf{T}_i - \mathbf{Z}\mathbf{S}\mathbf{S}^{-1}\mathbf{A}_i^\top \mathbf{T}_i\|^2, \tag{5.2}$$

where $\mathbf{G}_i$ $(n \times K_i)$ is an indicator matrix, $\mathbf{Q}_i$ $(K_i \times R_i)$ is a quantification matrix, and $\mathbf{T}_i$ $(R_i \times R_i)$ is an orthonormal matrix $(i = 1, \ldots, p)$. Murakami (1999) modified an orthomax criterion in three-way PCA for MCA solutions in which the loading matrix is rotated towards the simple structure by pre- and post-multiplied rotation matrices $\mathbf{S}$ and $\mathbf{T}_i$ $(i = 1, \ldots, p)$. In the algorithm, $\mathbf{S}$ and $\mathbf{T}_i$ $(i = 1, \ldots, p)$ are alternately optimized; the pre-multiplied rotation matrix $\mathbf{S}$ is optimized given the $\mathbf{T}_i$ $(i = 1, \ldots, p)$, and the post-multiplied matrices $\mathbf{T}_i$ $(i = 1, \ldots, p)$ is optimized given the matrix $\mathbf{S}$. It should be noted that the rotational freedom of the quantification matrices in nonlinear PCA disapears because all dimensions of the quantifications are set to be one.

5.3 Sparse MCA

In sparse MCA, there are mainly two approaches for obtaining the sparse solutions. In one approach, the MCA objective function is combined with a penalty function that penalizes the component weight matrix (Bernard et al. 2012). It should be mentioned that the usage of term "loading matrix" in this article is different to that of Bernard et al. (2012). The term "loading matrix" is used as the weights for the data matrix in Bernard et al. (2012). However, we call the loading matrix as the weights for the component score matrix and the component weight matrix as the weights for data matrix in this article. In the other approach, the loading matrix is estimated under the condition that the number of zeros are pre-specified, which is named pre-specified sparsity approach in this chapter. We propose the latter sparse MCA approach in this chapter because of the following two reasons. Firstly, the component weight matrix is focused rather than the loading matrix in the penalty approach. However, the loading matrix is practically used for interpretations of PCA or MCA solutions. Secondly, one of the drawbacks of the penalty-based approach is sometimes difficult to choose the appropriate value of the tuning parameter that defines the degree of the sparsity. In contrast, users can control the degree of sparsity in the loading matrix in the pre-specified sparsity approach.

The sparse MCA loss function is defined as

$$\sum_{i=1}^{p} \|\mathbf{G}_i \mathbf{Q}_i - \mathbf{Z}\mathbf{A}_i^\top\|^2, \tag{5.3}$$

subject to $card(\mathbf{A}) = v$ and ordinary MCA constraints $n^{-1}\mathbf{Z}^\top \mathbf{Z} = \mathbf{I}_r$, $\mathbf{J}\mathbf{Z} = \mathbf{Z}$, $\mathbf{1}_n^\top \mathbf{G}_i \mathbf{Q}_i = \mathbf{0}_{K_i}^\top$, and $n^{-1}\mathbf{Q}_i^\top \mathbf{G}_i^\top \mathbf{G}_i \mathbf{Q}_i = \mathbf{I}_{R_i}$. Here, $\mathbf{J}$ $(n \times n)$ is a centering matrix, and $card(\bullet) = v$ means that the number of zero elements (sparsity) is equal to v.

In the sparse PCA with pre-specifying sparsity approach, an efficient heuristic algorithm called greedy search is proposed to find component loadings sequentially with direct cardinality constraints (d'Aspremont et al. 2008; Moghaddam et al. 2006). Recently, an alternative algorithm has been proposed by Adachi and Trendafilov (2015), in which all component loadings are estimated with direct cardinality constraints. Adachi and Trendafilov (2015) pointed out that the simultaneous estimation in sparse PCA is superior to obtaining each component sequentially by the recovery of the true loading matrix. Thus, we adapted the Adachi and Trendafilov's algorithm and developed a simultaneous estimation method for sparse MCA.

The loss function (5.3) can be rewritten as

$$\sum_{i=1}^{p} \|\mathbf{G}_i\mathbf{Q}_i - \mathbf{Z}\mathbf{B}_i + \mathbf{Z}\mathbf{B}_i - \mathbf{Z}\mathbf{A}_i\|^2 = \sum_{i=1}^{p} \|\mathbf{G}_i\mathbf{Q}_i - \mathbf{Z}\mathbf{B}_i\|^2 + n\|\mathbf{B} - \mathbf{A}\|^2, \quad (5.4)$$

where $\mathbf{B} = n^{-1}\sum \mathbf{Q}_i^{\top}\mathbf{G}_i^{\top}\mathbf{Z}$. The loss function (5.4) can be solved by alternately performing three steps:

- *Q-step*: minimizing the loss function (5.4) over $\mathbf{Q}$ subject to $\mathbf{1}_n^{\top}\mathbf{G}_i\mathbf{Q}_i = \mathbf{0}_{K_i}^{\top}$, and $n^{-1}\mathbf{Q}_i^{\top}\mathbf{G}_i^{\top}\mathbf{G}_i\mathbf{Q}_i = \mathbf{I}_{R_i}$;
- *Z-step*: minimizing the loss function (5.4) over $\mathbf{Z}$ subject to $\mathbf{J}\mathbf{Z} = \mathbf{Z}$ and $n^{-1}\mathbf{Z}^{\top}\mathbf{Z} = \mathbf{I}_c$;
- *A-step*: minimizing the loss function (5.4) over $\mathbf{A}$ subject to $card(\mathbf{A}) = v$.

5.3.1 Q-Step

We first consider the Q-step, which is equivalent to the minimization of $\sum_{i=1}^{p}\|\mathbf{G}_i\mathbf{Q}_i - \mathbf{Z}\mathbf{B}_i\|^2$, under the restrictions $\mathbf{1}_n^{\top}\mathbf{G}_i\mathbf{Q}_i = \mathbf{0}_{K_i}^{\top}$, and $n^{-1}\mathbf{Q}_i^{\top}\mathbf{G}_i^{\top}\mathbf{G}_i\mathbf{Q}_i = \mathbf{I}_{R_i}$. $\sum_{i=1}^{p}\|\mathbf{G}_i\mathbf{Q}_i - \mathbf{Z}\mathbf{B}_i\|^2$ is expressed as $\sum_{i=1}^{p}\|\mathbf{G}_i\mathbf{D}_i^{-1/2}\mathbf{D}_i^{1/2}\mathbf{Q}_i - \mathbf{Z}\mathbf{B}_i\|^2$. Thus, the minimization for the quantification parameters reduces to the orthogonal Procrustes problem and the optimal $\mathbf{Q}_i$ can be obtained by the singular value decomposition (SVD) of $\mathbf{D}_i^{-1/2}\mathbf{G}_i^{\top}\mathbf{Z}\mathbf{B}^{\top}$ in the same manner as ordinary (rank-restricted) MCA in Sect. 3.3.

5.3.2 Z-Step

In Z-step, minimizing the loss function (5.4) equals the minimization of $\sum_{i=1}^{p}\|\mathbf{G}_i\mathbf{Q}_i - \mathbf{Z}\mathbf{B}_i\|^2$, under the restrictions $n^{-1}\mathbf{Z}^{\top}\mathbf{Z} = \mathbf{I}_r$, $\mathbf{J}\mathbf{Z} = \mathbf{Z}$ and can be regarded as the orthogonal Procrustes problem. Hence, we can have the optimal $\mathbf{Z}$ by the SVD of $\sum \mathbf{G}_i\mathbf{Q}_i\mathbf{B}_i$.

5.3.3 A-Step

For updating $\mathbf{A}$, we use the algorithm proposed by Adachi and Trendafilov (2015). The term $n\|\mathbf{B} - \mathbf{A}\|^2$ is only relevant to $\mathbf{A}$ in the loss function (5.4). Therefore, minimizing (5.4) is equivalent to the minimization of $n\|\mathbf{B}-\mathbf{A}\|^2$ subject to $card(A) = v$. Let us denote $f(\mathbf{A}) = \|\mathbf{B} - \mathbf{A}\|^2$. $f(\mathbf{A})$ is expressed as

$$f(\mathbf{A}) = \|\mathbf{B} - \mathbf{A}\|^2 = \sum_{(i,j)\in O} b_{ij}^2 + \sum_{(i,j)\in O^c} (b_{ij} - a_{ij})^2 \geq \sum_{(i,j)\in O} b_{ij}^2, \tag{5.5}$$

where O denotes the set of the $q = Rr - v$ indexes (i,j)'s indicating the locations where $\sum_{i=1}^{p} R_i = R$, the loadings a_{ij} are zero, and the compliment set O^c the $c(i,j)^s$ for nonzero a_{ij}. The inequality (5.5) shows that the function $f(\mathbf{A})$ attains its lower limit $\sum_{(i,j)\in O} b_{ij}^2$ when the non-zero loadings a_{ij} with $(i,j) \in O^c$ are set to equal to b_{ij}. Moreover, the limit $\sum_{(i,j)\in O} b_{ij}^2$ is minimal, when $\mathbf{O}$ contains the indexes for the q smallest b_{ij}^2 among all squared elements of $\mathbf{B}$. Hence, the optimal $\mathbf{A}$ can be obtained by

$$a_{ij} = \begin{cases} 0, & \textit{iff } b_{ij}^2 \leq b_{(q)}^2 \\ b_{ij}, & \textit{otherwise} \end{cases} \tag{5.6}$$

where $b_{(q)}^2$ is the q-th smallest value among all b_{ij}^2.

5.4 Real Data Example

The sparse MCA is illustrated using Japanese baseball data (Makino 2015) and compared with the varimax-rotated solution of ordinary MCA. This dataset describes the scores of 62 batters in Japanese professional baseball in 2010, and we use the following variables from the dataset: batting average, runs, doubles, home runs, runs batted in, and strikeouts (Table 5.1). We set the number of components to two and the dimension of quantification to one for all variables, which can be interpreted as expressing whether batters hit for average (table setters) or power (sluggers). Figure 5.1 illustrates the amount of variances for all possible zero cases. As shown in Fig. 5.1, the values are sharply changed when $card(A) = 7$. Thus, we use $card(A) = 6$ in this data analysis. The result of the loading matrix is reported in Table 5.2 and the rotated loading matrix of ordinary MCA solution is reported in Table 5.3. Similar loadings are obtained by sparse MCA and rotated MCA solution, but as you can see, the sparse MCA solution achieves the true simple structure.

Table 5.1 Japanese baseball data from Makino (2015)

Batting average	Run	Double	Home run	RBI	Strike
2	2	2	1	1	1
2	2	1	1	1	1
2	2	2	2	2	1
2	2	2	2	2	1
2	2	2	2	2	1
2	2	2	1	1	1
2	2	2	2	2	2
2	1	2	1	1	1
2	2	1	2	2	2
2	2	2	1	1	1
2	2	2	2	2	2
2	2	2	2	2	1
2	2	2	2	2	2
2	1	1	1	1	1
2	2	1	2	2	2
1	1	1	1	1	1
1	1	1	1	1	2
1	1	2	1	1	1
1	1	1	1	1	1
1	1	1	2	1	1
1	1	1	1	1	1
1	2	2	2	2	2
1	2	2	2	2	1
1	1	1	1	1	1
1	1	1	1	1	1
1	1	1	2	1	1
1	1	1	1	1	1
1	1	1	1	1	1
1	1	1	2	2	2
1	2	2	2	2	2
1	1	2	2	2	2
2	2	2	1	1	2
2	2	1	1	1	1

(continued)

Table 5.1 (continued)

Batting average	Run	Double	Home run	RBI	Strike
2	2	2	1	2	1
2	1	1	2	2	2
2	2	2	2	2	2
2	2	2	1	1	1
2	2	2	1	1	1
2	1	1	1	1	2
2	2	2	2	2	2
2	2	2	1	2	2
2	2	2	1	2	1
2	2	2	1	1	2
2	2	2	1	1	1
2	2	1	1	1	1
1	2	2	1	1	1
1	2	2	1	2	2
1	2	2	2	2	2
1	1	1	1	1	2
1	1	2	1	2	1
1	1	2	2	2	2
1	1	1	1	1	1
1	1	1	1	1	1
1	1	1	2	2	2
1	1	1	2	2	2
1	1	1	1	1	1
1	1	1	2	2	2
1	1	1	1	1	2
1	1	1	1	1	2
1	1	1	2	2	2
1	1	1	2	2	2
1	1	1	2	2	2

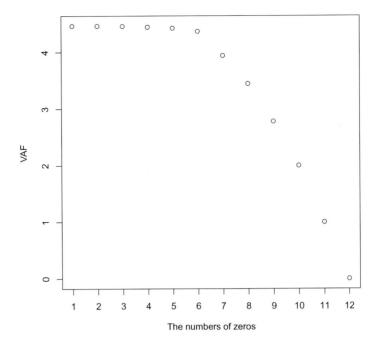

Fig. 5.1 The plot of VAF for number of zeros

Table 5.2 Component loadings in sparse MCA

	Component 1	Component 2
Batting average	**0.82**	0
Runs	**0.93**	0
Doubles	**0.81**	0
Home runs	0	**0.85**
Runs batted in	0	**0.90**
Strikeouts	0	**0.76**

Table 5.3 Rotated component loadings in ordinary MCA

	Component 1	Component 2
Batting average	**0.82**	0.1
Runs	**0.92**	−0.1
Doubles	**0.81**	−0.15
Home runs	0.02	**0.89**
Runs batted in	0.22	**0.90**
Strikeouts	−0.16	**0.76**

5.5 Discussion

There are mainly two approaches to estimate the sparse MCA solutions: the penalty-approach and pre-specifying sparsity approach. We introduced the latter approach to a sparse MCA method. The proposed method can be regarded as a simple extension of sparse PCA method proposed by Adachi and Trendafilov (2015). That is, when all observed variables are numerical in the introduced method, it is equal to the sparse PCA method proposed by Adachi and Trendafilov (2015). Thus, it can be extended to the case that observed data are a mixture of qualitative and quantitative variables.

Although we advocate the sparse MCA method with the pre-specifying sparsity approach, two problems are still present. First, the pre-specifying sparsity approach cannot be applied to high-dimensional datasets or purely exploratory situations about which the users do not have a hypothesis. The pre-specifying sparsity method is suitable for small datasets or users who hope to have a loading matrix with a specific number of zeros before the analysis. The penalty-based approach is appropriate for a situation in which the proposed method cannot be applied. Considering applications, a sparse MCA method with the penalty-based approach needs to be developed as well as the proposed method. Second, it is difficult to determine the optimal degree of sparsity. In this paper, we determined the number of zero elements by a screeplot of the amount of variance with a fixed the number of components. The application to large datasets and dimension setting of the degree of freedom, however, remains a subject for further discussion.

References

Adachi, K.: Oblique promax rotation applied to the solutions in multiple correspondence analysis. Behaviormetrika **31**, 1–12 (2004)

Adachi, K., Trendafilov, N.T.: Sparse principal component analysis subject to prespecified cardinality of loadings. Comput. Stat. **31**, 1–25 (2015)

Bernard, A., Guinot, C., Saporta, G.: Sparse principal component analysis for multiblock data and its extension to sparse multiple correspondence analysis. In: Colubi, A., Fokianos, K., Gonzalez-Rodriguez, G., Kontaghiorghes, E. (eds.) Proceedings of 20th International Conference on Computational Sttatistics (COMPSTAT 2012), pp. 99–106 (2012)

Browne, M.W.: An overview of analytic rotation in exploratory factor analysis. Multivar. Behav. Res. **36**, 111–150 (2001)

d'Aspremont, A., Bach, F., Ghaoui, L.E.: Optimal solutions for sparse principal component analysis. J. Mach. Learn. Res. **9**, 1269–1294 (2008)

Kiers, H.A.L.: Simple structure in component analysis techniques for mixtures of qualitative and quantitative variables. Psychometrika **56**, 197–212 (1991)

Makino, N.: Generalized data-fitting factor analysis with multiple quantification of categorical variables. Comput. Stat. **30**, 279–292 (2015)

Moghaddam, B., Weiss, Y., Avidan, S.: Spectral bounds for sparse PCA: exact and greedy algorithms. Adv. Neural Inf. Process. Syst. **18**, 915–922 (2006)

Murakami, T.: A psychometrics study on principal component analysis of categorical data, Technical report (in Japanese) (1999)

Murakami, T., Kiers, H.A.L., ten Berge, J.M.F.: Non-metric principal component analysis for categorical variables with multiple quantifications, Unpublished manuscript (1999)

Thurstone, L.L.: Multiple Factor Analysis. University of Chicago Press, Chicago (1947)

Trendafilov, N.T.: From simple structure to sparse components: a review. Comput. Stat. **29**, 431–454 (2014)

Trendafilov, N.T., Adachi, K.: Sparse vs simple structure loadings. Psychometrika **80**, 776–790 (2015)

van de Velden, M., Kiers, H.A.L.: An application of rotation in correspondence analysis, New Developments in Psychometrics, pp. 471–478 (2003)

Zou, H., Hastie, T., Tibshirani, R.: Sparse principal component analysis. J. Comput. Graph. Stat. **15**, 265–286 (2006)

Chapter 6
Joint Dimension Reduction and Clustering

Abstract Cluster analysis is a technique that attempts to divide objects into similar groups. As described in previous studies, cluster analysis works poorly when variables that do not reflect the clustering structure are present in the dataset or when the number of variables is large. In order to tackle this problem, several methods have been proposed that jointly perform clustering of objects and dimension reduction of the variables. In this chapter, we review the technique whereby multiple correspondence analysis and k-means clustering are combined in order to investigate the relationships between qualitative variables.

Keywords Clustering · Dimension reduction · Tandem analysis · k-means

6.1 Introduction

Cluster analysis, which is sometimes simply called clustering, is a widely used technique that attempts to divide objects into homogeneous groups. As noted in previous studies, cluster analysis does not work properly when variables that do not reflect the clustering structure are included in the dataset, or when the number of variables is large. In order to deal with this problem, researchers often perform dimension reduction of the variables using principal component analysis (PCA) or multiple correspondence analysis (MCA) before clustering the objects. However, clustering may still fail because the first few components do not necessarily define the subspace that is most informative about the cluster structure in the data (De Sorte and Carroll 1994). This two-step procedure is called tandem analysis. Several authors have argued against this approach and have instead proposed procedures in which dimension reduction of the variables and clustering of the objects are conducted simultaneously. In the case of numerical variables, a number of approaches have been developed for such simultaneous analysis, including reduced k-means (De Sorte and Carroll 1994), factorial k-means (Vichi and Kiers 2001), factor discriminant k-means (Rocci et al. 2011), and generalized reduced clustering (Yamamoto and Hwang 2014), in which PCA is used to summarize the variables and k-means (MacQueen 1967) is used for non-hierarchical clustering of the objects.

© The Author(s) 2016
Y. Mori et al., *Nonlinear Principal Component Analysis and Its Applications*,
JSS Research Series in Statistics, DOI 10.1007/978-981-10-0159-8_6

In the literature on qualitative variables, several methods have also been proposed. Van Buuren and Heiser (1989) proposed a method for clustering and optimal scaling of variables called GROUPALS, in which object scores are expressed as the product of the cluster membership and the cluster centroid matrices. In other techniques, Hwang et al. (2006) and Mitsuhiro and Yadohisa (2015) combined MCA and k-means clustering for simultaneous analysis in which cluster centroids and variable categories are jointly displayed in a low-dimensional space.

In this chapter, we focus on techniques that combine MCA and k-means for joint dimension reduction and clustering of qualitative variables. Note that a method that jointly performs dimension reduction and fuzzy clustering has also been proposed, but this is not the focus of this chapter (see Hwang et al. 2010 for details).

6.2   Simultaneous Analysis with MCA and k-Means

6.2.1   Hwang, Dillon and Takane's Method

Let $\mathbf{G}_i$ be an n-objects $\times$ K_i-categories matrix of dummy-coded categorical variables, $\mathbf{W}_i$ be a K_i-categories $\times$ r-components matrix of weights, or category coordinate, $\mathbf{Z}$ be an n-objects $\times$ r-components matrix of component score, $\mathbf{U}$ be an n-objects $\times$ t-clusters matrix of indicator variables, allocating objects into one of t clusters, and $\mathbf{\Gamma}$ be an t-clusters $\times$ r-component matrix of the centroid of clusters. We also denote α_1 and α_2 as non-negative scalar weights. In Hwang et al. (2006), the loss function was defined as

$$\alpha_1 \sum_{i=1}^{p} \|\mathbf{Z} - \mathbf{G}_i \mathbf{W}_i\|^2 + \alpha_2 \|\mathbf{Z} - \mathbf{U}\mathbf{\Gamma}\|^2, \tag{6.1}$$

and minimized over $\mathbf{Z}$, $\mathbf{\Gamma}$, $\mathbf{W}_i$ and $\mathbf{U}$ ($i = 1, \ldots, p$), subject to $\mathbf{Z}^\top \mathbf{Z} = \mathbf{I}_r$ and $\alpha_1 + \alpha_2 = 1$. Here p indicates the number of variables, and α_1 and α_2 are specified by practitioners before the analysis. It is obvious that the loss function (6.1) reduces to the standard homogeneity analysis criterion for MCA (Gifi 1990) if $\alpha_1 = 1$, and to the standard k-means criterion if $\alpha_2 = 1$. Hwang et al. (2006) has discussed the role of the values α_1 and α_2, designed to balance out the two terms for MCA and k-means, when specifying $\alpha_1 = \alpha_2 = 0.5$. If the practitioners judges that classification is more important than dimension reduction, the weights are set to $\alpha_1 < \alpha 2$, and vice versa.

An alternating least squares algorithm was developed to minimize the loss function (6.1), which comprises three main steps. First, $\mathbf{Z}$ is updated for fixed $\mathbf{W}_i$, $\mathbf{U}$ and $\mathbf{\Gamma}$. The loss function (6.1) is expanded as

$$(\alpha_1 p + \alpha_2)c - \mathrm{tr}\left(\mathbf{Z}^\top \left[\alpha_1 \sum_{i=1}^{p} \mathbf{P}_G + \alpha_2 \mathbf{P}_U\right] \mathbf{Z}\right), \tag{6.2}$$

where $\mathbf{P}_G = \mathbf{G}_i(\mathbf{G}_i^\top \mathbf{G}_i)^{-1}\mathbf{G}_i^\top$ is the projection matrix of $\mathbf{G}_i$, and $\mathbf{P}_U = \mathbf{U}(\mathbf{U}^\top \mathbf{U})^{-1}\mathbf{U}^\top$ is the projection matrix of $\mathbf{U}$. The minimization of the loss function (6.1) with respect to $\mathbf{Z}$ is equivalent to the maximization of (6.2). This optimization can be achieved by the eigenvalue decomposition of $\alpha_1 \sum_{i=1}^{p} \mathbf{P}_{G_i} + \alpha_2 \mathbf{P}_U$ (e.g., ten Berge 1993). Second, $\mathbf{U}$ is updated for fixed $\mathbf{W}_i$, $\mathbf{Z}$ and $\mathbf{\Gamma}$. This amounts to minimizing the second term of the loss function (6.1) over $\mathbf{U}$, thus we use the ordinary k-means algorithm to minimize this criterion. Finally, $\mathbf{W}_i$ and $\mathbf{\Gamma}$ are updated for fixed $\mathbf{Z}$ and $\mathbf{U}$. We can obtain $\mathbf{W}_i$ and $\mathbf{\Gamma}$ for solving the regression problem, because $\mathbf{W}_i$ is involved only in the first term and $\mathbf{\Gamma}$ only in the second term of the loss function (6.1). Thus, their updates are given by

$$\mathbf{W}_i = (\mathbf{G}_i^\top \mathbf{G}_i)^{-1}\mathbf{G}_i^\top \mathbf{Z}, \tag{6.3}$$

$$\mathbf{\Gamma} = (\mathbf{U}^\top \mathbf{U})^{-1}\mathbf{U}^\top \mathbf{Z}. \tag{6.4}$$

The algorithm is summarized as follows:

Step 1: Choose initial values for $\mathbf{Z}$, and $\mathbf{U}$. Then, $\mathbf{W}_i$ and $\mathbf{\Gamma}$ are given by $\mathbf{W}_i = (\mathbf{G}_i^\top \mathbf{G}_i)^{-1}\mathbf{G}_i^\top \mathbf{Z}$ and $\mathbf{\Gamma} = (\mathbf{U}^\top \mathbf{U})^{-1}\mathbf{U}^\top \mathbf{Z}$.

Step 2: Update the component score matrix $\mathbf{Z}$ by the eigenvalue decomposition of $\alpha_1 \sum_{i=1}^{p} \mathbf{P}_{G_i} + \alpha_2 \mathbf{P}_U$.

Step 3: Update the membership matrix $\mathbf{U}$ using the ordinary k-means algorithm.

Step 4: Update the category coordinate matrices $\mathbf{W}_i$ for each variable and the cluster centroid matrix $\mathbf{\Gamma}$ by $\mathbf{W}_i = (\mathbf{G}_i^\top \mathbf{G}_i)^{-1}\mathbf{G}_i^\top \mathbf{Z}$ and $\mathbf{\Gamma} = (\mathbf{U}^\top \mathbf{U})^{-1}\mathbf{U}^\top \mathbf{Z}$.

Step 5: Terminate if the difference between the values of the loss function (6.1) in the current and previous steps is less than ε; otherwise, return to *Step 2*.

6.2.2 Mitsuoka and Yadohisa's Method

Let $\mathbf{Q}_i$ be a K_i-categories $\times$ R_i-dimensions matrix of quantification parameters, $\mathbf{J}$ be an n-objects $\times$ n-objects of centering matrix, and $\mathbf{A}_i$ be a R_i-dimensions $\times$ r-components matrix of the loading matrix of the ith variable.

Then, the loss function of the Mitsuhiro and Yadohisa (2015) method is expressed as

$$\sum_{i=1}^{p} \|\mathbf{G}_i \mathbf{Q}_i - \mathbf{Z}\mathbf{A}_i^\top\|^2 + \|\mathbf{Z} - \mathbf{U}\mathbf{\Gamma}\|^2, \tag{6.5}$$

and minimized over $\mathbf{Z}$, $\mathbf{U}$, $\boldsymbol{\Gamma}$, $\mathbf{A}_i$ and $\mathbf{Q}_i$ $(i = 1, \ldots, p)$, subject to $n^{-1}\mathbf{Z}^{\top}\mathbf{Z} = \mathbf{I}_r$, $\mathbf{JZ} = \mathbf{Z}$, $\mathbf{1}_n^{\top}\mathbf{G}_i\mathbf{Q}_i = \mathbf{0}_{K_i}^{\top}$, and $n^{-1}\mathbf{Q}_i^{\top}\mathbf{G}_i^{\top}\mathbf{G}_i\mathbf{Q}_i = \mathbf{I}_{R_i}$. Comparing the loss function (6.5) with (6.1), the first terms that define the MCA criterion are different. MCA is formulated by the homogeneity analysis criterion (Gifi 1990) in Hwang et al. (2006), whereas in Mitsuhiro and Yadohisa (2015), MCA is formulated as approximating a data matrix by a lower rank matrix using the quantification technique (Murakami et al. 1999; Murakami 1999; Adachi and Murakami 2011). Please see Chap. 2 for details of the latter MCA formulation.

The parameters $\mathbf{Z}$, $\mathbf{U}$, $\boldsymbol{\Gamma}$, $\mathbf{A}_i$ and $\mathbf{Q}_i$ $(i = 1, \ldots, p)$ can be estimated by alternately updating until convergence. In the first step, $\mathbf{Q}_i$ is updated for fixed the other parameters. Re-expressing the first term of the loss function (6.5) as

$$\mathbf{G}_i\mathbf{Q}_i = \mathbf{G}_i(\mathbf{G}_i^{\top}\mathbf{G}_i)^{-1/2}(\mathbf{G}_i^{\top}\mathbf{G}_i)^{1/2}\mathbf{Q}_i, \tag{6.6}$$

$$= \mathbf{B}_i\boldsymbol{\Lambda}_i, \tag{6.7}$$

$\mathbf{Q}_i$ can be updated by solving the orthogonal Procrustes problem for $\boldsymbol{\Lambda}_i$, because $\mathbf{Q}_i$ is involved only in the first term and $\boldsymbol{\Lambda}_i$ is a column-orthogonal matrix. In the second step, $\mathbf{Z}$ is updated for fixed the other parameters. We can express this loss function (6.5) as

$$const - \mathrm{tr}([\mathbf{B}_Q^{\top}\mathbf{G}^{\top} + \boldsymbol{\Gamma}^{\top}\mathbf{U}^{\top}]\mathbf{J})\mathbf{Z}, \tag{6.8}$$

where $const$ denotes a constant that is irrelevant to the parameters and $\mathbf{B}_Q$ is the block diagonal matrix of $\mathbf{Q}_i$ $(i = 1, \ldots, p)$. The minimization of the loss function (6.5) with respect to $\mathbf{Z}$ is equivalent to the maximization of (6.8). This optimization can be achieved by the singular value decomposition of $(\mathbf{B}_Q^{\top}\mathbf{G}^{\top} + \boldsymbol{\Gamma}^{\top}\mathbf{U}^{\top})\mathbf{J}$ (e.g., ten Berge 1993). In the third step, we update $\mathbf{U}$ for fixed the other parameters. This amounts to minimizing the second term of the loss function (6.5) over $\mathbf{U}$, and thus we update $\mathbf{U}$ by using the ordinary k-means algorithm. In the final step, we update $\mathbf{A}_i$ and $\boldsymbol{\Gamma}$ for fixed $\mathbf{Q}_i$, $\mathbf{Z}$, and $\mathbf{U}$. We can update for $\mathbf{A}_i$ and $\boldsymbol{\Gamma}$ by solving the following regression problems:

$$\mathbf{A}_i = n^{-1}\mathbf{Q}_i^{\top}\mathbf{G}_i^{\top}\mathbf{Z}, \tag{6.9}$$

$$\boldsymbol{\Gamma} = (\mathbf{U}^{\top}\mathbf{U})^{-1}\mathbf{U}^{\top}\mathbf{Z}, \tag{6.10}$$

because $\mathbf{A}_i$ is involved only in the first term and $\boldsymbol{\Gamma}$ is only in the second term of the loss function (6.5).

The algorithm is summarized as follows:

Step 1: Choose initial values for $\mathbf{Z}$, $\mathbf{A}_i$, $\mathbf{Q}_i$, $\mathbf{U}$, and $\boldsymbol{\Gamma}$. Take arbitrary matrices $\mathbf{Q}_i$, $\mathbf{Z}$, and $\mathbf{U}$ which satisfies the constraints. Then, $\mathbf{A}_i$ and $\boldsymbol{\Gamma}$ are given as $\mathbf{A}_i = n^{-1}\mathbf{Q}_i^{\top}\mathbf{G}_i^{\top}\mathbf{Z}$ for each variable and $\boldsymbol{\Gamma} = (\mathbf{U}^{\top}\mathbf{U})^{-1}\mathbf{U}^{\top}\mathbf{Z}$.

Step 2: Update the quantification parameters by solving the orthogonal Procrustes problems.

Step 3: Update the component score matrix $\mathbf{Z}$ by the singular value decomposition of $(\mathbf{B}_Q^{\top}\mathbf{G}^{\top} + \mathbf{\Gamma}^{\top}\mathbf{U}^{\top})\mathbf{J}$.

Step 4: Update the membership matrix $\mathbf{U}$ using the ordinary k-means algorithm.

Step 5: Update the loading and cluster centroid matrices by $\mathbf{A}_i$ and $\mathbf{\Gamma}$, given by $\mathbf{A}_i = n^{-1}\mathbf{Q}_i^{\top}\mathbf{G}_i^{\top}\mathbf{Z}$ for each variable and $\mathbf{\Gamma} = (\mathbf{U}^{\top}\mathbf{U})^{-1}\mathbf{U}^{\top}\mathbf{Z}$.

Step 6: Terminate if the difference between the values of the loss function (6.5) in the current and previous steps is less than ε; otherwise, return to *Step 2*.

When the optimal parameters $\mathbf{Z}$, $\mathbf{U}$, $\mathbf{\Gamma}$, $\mathbf{A}_i$ and $\mathbf{Q}_i$ $(i = 1, \ldots, p)$ have been obtained, we have the category coordinate matrix $\mathbf{W}_i = \mathbf{Q}_i\mathbf{A}_i$ for each variable. Mitsuhiro and Yadohisa (2015) pointed out that unlike in Hwang et al. (2006), not only the variable categories and cluster centroids are displayed but also the object scores. This is because the object scores estimated as vector of ones that have no meaning in the first dimension can be removed, and we can interpret the relationship between objects and categories at each cluster and extract features from each cluster. However, the abovementioned deficit can be removed in Hwang et al. (2006) as discussed in the next section.

6.3 Differences Between Hwang, Dillon and Takane's Method and Mitsuoka and Yadohisa's Method

Hwang et al. (2006) and Mitsuhiro and Yadohisa (2015) introduced joint dimension reduction and clustering methods for the qualitative variables. We discuss the relationship between them in more detail in this section.

Relation

The loss function (6.1) and (6.5) are equivalent when $\alpha_1 = \alpha_2 = 0.5$ and $\mathbf{JZ} = \mathbf{Z}$ are imposed in (6.1).

The relationship can be shown by expanding the loss functions (6.1) and (6.5). Under the constraints $\alpha_1 = \alpha_2 = 0.5$ and $\mathbf{JZ} = \mathbf{Z}$, the loss function (6.1) can be expressed as

$$0.5\left\{\sum_{i=1}^{p}\|\mathbf{Z} - \mathbf{G}_i\mathbf{W}_i\|^2 + \|\mathbf{Z} - \mathbf{U}\mathbf{\Gamma}\|^2\right\}$$

$$= 0.5\left\{nrp - 2\mathrm{tr}\left(\sum_{i=1}^{p}\mathbf{F}^{\top}\mathbf{G}_i\mathbf{W}_i\right) + \mathrm{tr}\left(\sum_{i=1}^{p}\mathbf{W}_i^{\top}\mathbf{G}_i^{\top}\mathbf{G}_i\mathbf{W}_i\right) + \|\mathbf{Z} - \mathbf{U}\mathbf{\Gamma}\|^2\right\},$$

$$(6.11)$$

and the loss function (6.5) as

$$\sum_{i=1}^{p} \|\mathbf{G}_i\mathbf{Q}_i - \mathbf{Z}\mathbf{A}_i^\top\|^2 + \|\mathbf{Z} - \mathbf{U}\mathbf{\Gamma}\|^2$$

$$= npR + -2\text{tr}\left(\sum_{i=1}^{p} \mathbf{Z}^\top\mathbf{G}_i\mathbf{Q}_i\mathbf{A}\right) + n * \text{tr}\left(\sum_{i=1}^{p} \mathbf{A}_i^\top\mathbf{A}\right) + \|\mathbf{Z} - \mathbf{U}\mathbf{\Gamma}\|^2$$

$$= npR + -2\text{tr}\left(\sum_{i=1}^{p} \mathbf{Z}^\top\mathbf{G}_i\mathbf{Q}_i\mathbf{A}\right) + \text{tr}\left(\sum_{i=1}^{p} \mathbf{A}_i^\top\mathbf{Q}_i^\top\mathbf{G}_i^\top\mathbf{G}_i\mathbf{Q}_i\mathbf{A}\right) + \|\mathbf{Z} - \mathbf{U}\mathbf{\Gamma}\|^2$$

$$= npR + -2\text{tr}\left(\sum_{i=1}^{p} \mathbf{F}^\top\mathbf{G}_i\mathbf{W}_i\right) + \text{tr}\left(\sum_{i=1}^{p} \mathbf{W}_i^\top\mathbf{G}_i^\top\mathbf{G}_i\mathbf{W}_i\right) + \|\mathbf{Z} - \mathbf{U}\mathbf{\Gamma}\|^2,$$

$$(6.12)$$

where $\sum(R_i) = R$. As the scale 0.5 is irrelevant to the parameters in the loss function (6.11), Hwang et al. (2006) and Mitsuhiro and Yadohisa (2015) are essentially equivalent. Adding the constraint $\mathbf{JZ} = \mathbf{Z}$ discards the trivial solution in Hwang et al. (2006).

The relationship above is also suggests that if we modify the loss function (6.5) as

$$\alpha_1 \sum_{i=1}^{p} \|\mathbf{G}_i\mathbf{Q}_i - \mathbf{Z}\mathbf{A}_i^\top\|^2 + \alpha_2\|\mathbf{Z} - \mathbf{U}\mathbf{\Gamma}\|^2, \qquad (6.13)$$

with constraints $n^{-1}\mathbf{Z}^\top\mathbf{Z} = \mathbf{I}_r$, $\mathbf{JZ} = \mathbf{Z}$, $\mathbf{1}_n^\top\mathbf{G}_i\mathbf{Q}_i = \mathbf{0}_{K_i}^\top$, $n^{-1}\mathbf{Q}_i^\top\mathbf{G}_i^\top\mathbf{G}_i\mathbf{Q}_i = \mathbf{I}_{R_i}$, and $\alpha_1 + \alpha_2 = 1$, it is equivalent to the method of Hwang et al. (2006).

6.4   Real Data Example

We demonstrated joint dimension reduction and clustering using South Korean underwear data (Hwang et al. 2006). The Mitsuhiro and Yadohisa (2015) method was used for this demonstration. In a survey, 664 South Korean consumers were asked to provide responses to three multiple choice items: preferred brand of underwear (8 brands), attributes when purchasing a brand of underwear (15 attributes) and consumer age (6 levels). The dataset can be obtained from cluster package in R (Markos et al. 2015). We choose $r = 2$, $t = 3$ and the dimension of quantification is set to be two for all variables following Hwang et al. (2006). The estimated result is shown in Fig. 6.1, in which the variable categories, the object scores and cluster centroids are displayed.

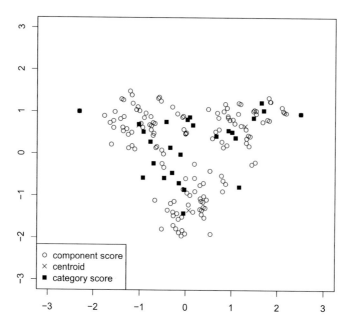

Fig. 6.1 The plot of component scores, cluster centroids, and category coordinates

References

Adachi, K., Murakami, T.: Nonmetric Multivariate Analysis: MCA, NPCA, and PCA. Asakura Shoten, Tokyo (2011). (in Japanese)

De Sorte, G., Carroll, J.D.: K-means clustering in a low-dimensional Euclidean space. In: Diday, E., Lechevallier, Y., Schader, M., Bertrand, P., Burtschy, B. (eds.) New Approached in Classification and Data Analysis, pp. 212-219. Springer, Berlin (1994)

Gifi, A.: Nonlinear Multivariate Analysis. Wiley, Chichester (1990)

Hwang, H., Dillon, W.R., Takane, Y.: An extension of multiple correspondence analysis for identifying heterogeneous subgroups of respondents. Psychometrika **71**, 61–171 (2006)

Hwang, H., Dillon, W.R., Takane, Y.: Fuzzy cluster multiple correspondence analysis. Behaviormetrika **37**, 111–133 (2010)

MacQueen, J.: Some methods for classification and analysis of multivariate observations. Proc. Fifth Berkeley Symp. Math. Stat. Probab. **1**, 281–297 (1967)

Markos, A., Iodice D' Enza, A., Van de Velden: Package clustrd (2015)

Mitsuhiro, M., Yadohisam, H.: Reduced k-means clustering with MCA in a low-dimensional space. Comput. Stat. **30**, 463–475 (2015)

Murakami, T.: A psychometrics study on principal component analysis of categorical data, Technical report (1999). (in Japanese)

Murakami, T., Kiers, H.A.L., ten Berge, J.M.F.: Non-metric principal component analysis for categorical variables with multiple quantifications (1999) (Unpublished manuscript)

Rocci, R., Garrone, S.A., Vichi, M.: A new dimension reduction method: factor discriminant k-means. J. Classif. **28**, 210–226 (2011)

ten Berge, J.M.F.: Least Squares Optimazation in Multivariate Analysis. DSNO Press, Leiden (1993)

Van Buuren, S., Heiser, W.J.: Clustering N objects into K groups under optimal scaling of variables. Psychometrika **54**, 699–706 (1989)

Vichi, M., Kiers, H.A.L.: Factorial k-means analysis for two-way data. Comput. Stat. Data Anal. **37**, 49–64 (2001)

Yamamoto, M., Hwang, H.: A general formulation of cluster analysis with dimension reduction and subspace separation. Behaviormetrika **41**, 115–129 (2014)

Chapter 7
Acceleration of Convergence
of the Alternating Least Squares Algorithm
for Nonlinear Principal Component Analysis

Abstract Nonlinear principal component analysis (PCA) requires iterative computation using the alternating least squares (ALS) algorithm, which alternates between optimal scaling for quantifying qualitative data and the analysis of optimally scaled data using the ordinary PCA approach. PRINCIPALS of Young et al. (Psychometrika 43:279–281) (1978) and PRINCALS of Gifi (Nonlinear Multivariate Analysis. Wiley, Chichester) (1990) are the ALS algorithms used for nonlinear PCA. When applying nonlinear PCA to very large data sets of numerous nominal and ordinal variables, the ALS algorithm may require many iterations and significant computation time to converge. One reason for the slow convergence of the ALS algorithm is that the speed of convergence is linear. In order to accelerate the convergence of the ALS algorithm, Kuroda et al. (Comput Stat Data Anal 55:143–153) (2011) developed a new iterative algorithm using the vector ε algorithm by Wynn (Math Comput 16:301–322) (1962).

Keywords Alternating least squares algorithm · Vector ε algorithm

7.1 Brief Introduction to the ALS Algorithm for Nonlinear PCA

Two versions of the alternating least squares (ALS) algorithm are used for the computation of nonlinear principal component analysis (PCA): PRINCIPALS Young et al. (1978) and PRINCALS Gifi (1990). For mixed measurement level data $\mathbf{Y}$ given by an $n \times p$ matrix, PRINCALS alternates between estimating $\mathbf{Z}$ and $\mathbf{A}$ in ordinary PCA and updating $\mathbf{Y}^*$ in quantifying $\mathbf{Y}$, where $\mathbf{Z}$ is an $n \times r$ matrix of n component scores on r $(1 \leq r \leq p)$ components, $\mathbf{A}$ is a $p \times r$ matrix of p component loadings on r components, and $\mathbf{Y}^*$ is an $n \times p$ optimally scaled matrix of $\mathbf{Y}$. PRINCALS estimates category quantification matrix $\mathbf{W}$ and updates $\mathbf{Z}$ using $\mathbf{W}$. Then, $\mathbf{Y}^*$ and $\mathbf{A}$ can be computed from $\mathbf{W}$. These algorithms are derived in Chap. 2.

Kuroda et al. (2011) derived an accelerated ALS algorithm for nonlinear PCA. The proposed algorithm speeds up the convergence of the ALS sequence via the vector ε (vε) algorithm by Wynn (1962). They demonstrated that the speed of convergence

© The Author(s) 2016
Y. Mori et al., *Nonlinear Principal Component Analysis and Its Applications*,
JSS Research Series in Statistics, DOI 10.1007/978-981-10-0159-8_7

of the acceleration algorithm is significantly faster than that of the ALS algorithm. The $v\varepsilon$ accelerated ALS algorithm is referred to as the $v\varepsilon$-ALS algorithm.

7.2   The Vector ε Algorithm

We introduce the $v\varepsilon$ algorithm of Wynn (1962) used in the acceleration of the ALS algorithm. The $v\varepsilon$ algorithm is utilized to speed up the convergence of a slowly convergent vector sequence and is very effective for linearly converging sequences.

Let $\mathbf{x}^{(t)}$ denote a vector of dimensionality d that converges to a vector $\mathbf{x}^{(\infty)}$ as $t \to \infty$. Let the inverse $[\mathbf{x}]^{-1}$ of a vector $\mathbf{x}$ be defined by

$$[\mathbf{x}]^{-1} = \frac{\mathbf{x}}{\|\mathbf{x}\|^2},$$

where $\|\mathbf{x}\|$ is the Euclidean norm of $\mathbf{x}$.

In general, the $v\varepsilon$ algorithm for a sequence $\{\mathbf{x}^{(t)}\}_{t \geq 0}$ starts with

$$\varepsilon^{(t,-1)} = 0, \qquad \varepsilon^{(t,0)} = \mathbf{x}^{(t)},$$

and then generates a vector $\varepsilon^{(t,k+1)}$ by

$$\varepsilon^{(t,k+1)} = \varepsilon^{(t+1,k-1)} + \left[\Delta\varepsilon^{(t,k)}\right]^{-1}, \qquad k = 0, 1, 2, \ldots, \tag{7.1}$$

where $\Delta\varepsilon^{(t,k)} = \varepsilon^{(t+1,k)} - \varepsilon^{(t,k)}$. In order to obtain $\varepsilon^{(t,k)}$, we require to compute $\varepsilon^{(t,0)}, \varepsilon^{(t,1)}, \ldots, \varepsilon^{(t,k-1)}$ recursively. Then the computational cost of $\varepsilon^{(t,k)}$ for a larger k is expensive. For practical implementation, we apply the $v\varepsilon$ algorithm for $k = 1$ to accelerate the convergence of $\{\mathbf{x}^{(t)}\}_{t \geq 0}$. We have, from Eq. (7.1),

$$\varepsilon^{(t,1)} = \varepsilon^{(t+1,-1)} + \left[\Delta\varepsilon^{(t,0)}\right]^{-1} = \left[\Delta\varepsilon^{(t,0)}\right]^{-1},$$
$$\varepsilon^{(t,2)} = \varepsilon^{(t+1,0)} + \left[\Delta\varepsilon^{(t,1)}\right]^{-1},$$

and then obtain

$$\varepsilon^{(t,2)} = \varepsilon^{(t+1,0)} + \left[\left[\Delta\varepsilon^{(t+1,0)}\right]^{-1} - \left[\Delta\varepsilon^{(t,0)}\right]^{-1}\right]^{-1}$$
$$= \mathbf{x}^{(t+1)} + \left[\left[\Delta\mathbf{x}^{(t+1)}\right]^{-1} - \left[\Delta\mathbf{x}^{(t)}\right]^{-1}\right]^{-1}.$$

We denote $\varepsilon^{(t,2)}$ as $\dot{\mathbf{x}}^{(t)}$.

Assume that $\{\mathbf{x}^{(t)}\}_{t \geq 0}$ converges to a vector $\mathbf{x}^{(\infty)}$. Then, in many cases, $\{\dot{\mathbf{x}}^{(t)}\}_{t \geq 0}$ generated by the $v\varepsilon$ algorithm converges to $\mathbf{x}^{(\infty)}$ faster than $\{\mathbf{x}^{(t)}\}_{t \geq 0}$.

Note that, at each iteration, the vε algorithm requires only $O(d)$ arithmetic operations while the Newton-Raphson algorithm is achieved at $O(d^3)$ and thus the computational cost is likely to increase as d becomes large.

7.3 Acceleration of Convergence of the ALS Algorithm for Nonlinear PCA

We assume that $\{\mathbf{Y}^{*(t)}\}_{t\geq 0}$ generated by the ALS algorithm converges to a vector $\mathbf{Y}^{*(\infty)}$. Then vε algorithm produces a faster convergent sequence $\{\dot{\mathbf{Y}}^{*(t)}\}_{t\geq 0}$ of $\{\mathbf{Y}^{*(t)}\}_{t\geq 0}$ and enables the acceleration of convergence of the ALS algorithm. We show the vε acceleration algorithm for PRINCIPALS and PRINCALS.

The vε acceleration for PRINCIPALS alternates the ALS and Acceleration steps:

The vε-ALS Algorithm: PRINCIPALS

- *ALS step*: Obtain $\mathbf{Y}^{*(t+1)}$ using PRINCIPALS:

  - *Model estimation step*: By solving the eigen-decomposition of $\mathbf{Y}^{*(t)\top}\mathbf{Y}^{*(t)}/n$ or the singular value decomposition of $\mathbf{Y}^{*(t)}$, obtain $\mathbf{A}^{(t+1)}$ and compute $\mathbf{Z}^{(t+1)} = \mathbf{Y}^{*(t)}\mathbf{A}^{(t+1)}$. Update $\hat{\mathbf{Y}}^{(t+1)} = \mathbf{Z}^{(t+1)}\mathbf{A}^{(t+1)\top}$.
  - *Optimal scaling step*: Obtain $\mathbf{Y}^{*(t+1)}$ by separately estimating $\mathbf{y}_j^*$ for each variable j. Compute $\mathbf{q}_j^{(t+1)}$ for nominal variables as

  $$\mathbf{q}_j^{(t+1)} = (\mathbf{G}_j^\top \mathbf{G}_j)^{-1} \mathbf{G}_j^\top \hat{\mathbf{y}}_j^{(t+1)}.$$

  Re-compute $\mathbf{q}_j^{(t+1)}$ for ordinal variables using the monotone regression (Kruskal 1964). For nominal and ordinal variables, update $\mathbf{y}_j^{*(t+1)} = \mathbf{G}_j \mathbf{q}_j^{(t+1)}$ and standardize $\mathbf{y}_j^{*(t+1)}$. For numerical variables, standardize observed vector $\mathbf{y}_j$ and set $\mathbf{y}_j^{*(t+1)} = \mathbf{y}_j$.

- *Acceleration step*: Calculate $\dot{\mathbf{Y}}^{*(t-1)}$ from $\{\mathbf{Y}^{*(t-1)}, \mathbf{Y}^{*(t)}, \mathbf{Y}^{*(t+1)}\}$ using the vε algorithm:

  $$\text{vec}\dot{\mathbf{Y}}^{*(t-1)} = \text{vec}\mathbf{Y}^{*(t)} + \left[\left[\Delta\text{vec}\mathbf{Y}^{*(t)}\right]^{-1} - \left[\Delta\text{vec}\mathbf{Y}^{*(t-1)}\right]^{-1}\right]^{-1},$$

  where $\text{vec}\mathbf{Y}^* = (\mathbf{y}_1^{*\top}\ \mathbf{y}_2^{*\top}\cdots\ \mathbf{y}_p^{*\top})^\top$, and check the convergence by

  $$\left\|\text{vec}\Delta\dot{\mathbf{Y}}^{*(t-2)}\right\|^2 < \delta,$$

  where δ is a desired accuracy.

Because the vε-ALS algorithm is designed to generate $\{\dot{\mathbf{Y}}^{*(t)}\}_{t\geq 0}$ converging to $\mathbf{Y}^{*(\infty)}$, the estimate of $\mathbf{Y}^*$ can be obtained from the final value of $\{\dot{\mathbf{Y}}^{*(t)}\}_{t\geq 0}$ after

the termination of the algorithm. The estimates of $\mathbf{Z}$ and $\mathbf{A}$ can then be calculated immediately from the estimate of $\mathbf{Y}^*$ in the Model parameter estimation step of the ALS step.

The $v\varepsilon$-ALS algorithm generates $\{\dot{\mathbf{Y}}^{*(t-1)}\}_{t\geq 0}$ and $\{\mathbf{Y}^{*(t+1)}\}_{t\geq 0}$ independently and does not use a vector $\dot{\mathbf{Y}}^{*(t-1)}$ as the estimate $\mathbf{Y}^{*(t+1)}$ in the next ALS step. Thus the algorithm speeds up the convergence of $\{\mathbf{Y}^{*(t)}\}_{t\geq 0}$ without affecting the convergence properties of the ordinary ALS algorithm.

We give the $v\varepsilon$-ALS algorithm using PRINCALS as the ALS algorithm. The acceleration algorithm speeds up the convergence of $\{\mathbf{Z}^{(t)}\}_{t\geq 0}$ by generating the fast converging sequence $\{\dot{\mathbf{Z}}^{(t)}\}_{t\geq 0}$ of $\{\mathbf{Z}^{(t)}\}_{t\geq 0}$.

The $v\varepsilon$-ALS Algorithm: PRINCALS

- *ALS step*: Obtain $\mathbf{Z}^{(t+1)}$ using PRINCALS:

  – *Estimation of category quantifications:* Compute $\mathbf{W}_j^{(t+1)}$ for $j = 1, \ldots, p$ as

  $$\mathbf{W}_j^{(t+1)} = (\mathbf{G}_j^\top \mathbf{G}_j)^{-1} \mathbf{G}_j^\top \mathbf{Z}^{(t)}.$$

  For the multiple variables in $\mathbf{J}_M$, set $\mathbf{W}_j^{(t+1)}$ to the estimate of multiple category quantifications.
  For the single variables in $\mathbf{J}_S$, update $\mathbf{a}_j^{(t+1)}$ by

  $$\mathbf{a}_j^{(t+1)\top} = \mathbf{W}_j^{(t+1)\top}(\mathbf{G}_j^\top \mathbf{G}_j)\mathbf{q}_j^{(t)} \Big/ \mathbf{q}_j^{(t)\top}(\mathbf{G}_j^\top \mathbf{G}_j)\mathbf{q}_j^{(t)}$$

  and compute $\mathbf{q}_j^{(t+1)}$ for nominal variables by

  $$\mathbf{q}_j^{(t+1)} = \mathbf{W}_j^{(t+1)}\mathbf{a}_j^{(t+1)\top} \Big/ \mathbf{a}_j^{(t+1)}\mathbf{a}_j^{(t+1)\top}.$$

  Re-compute $\mathbf{q}_j^{(t+1)}$ for ordinal variables using the monotone regression in a similar manner of PRINCIPALS. For numerical variables, standardize observed vector $\mathbf{y}_j$ and compute $\mathbf{q}_j^{(t+1)} = (\mathbf{G}_j^\top \mathbf{G}_j)^{-1}\mathbf{G}_j^\top \mathbf{y}_j$. Update $\mathbf{W}_j^{(t+1)} = \mathbf{q}_j^{(t+1)}\mathbf{a}_j^{(t+1)}$ for ordinal and numerical variables.
  – *Update of object scores:* Compute $\mathbf{Z}^{(t+1)}$ by

  $$\mathbf{Z}^{(t+1)} = \frac{1}{p}\sum_{j=1}^{p}\mathbf{G}_j\mathbf{W}_j^{(t+1)}.$$

  Column-wise center and orthonormalize $\mathbf{Z}^{(t+1)}$.

- *Acceleration step*: Calculate $\dot{\mathbf{Z}}^{(t-1)}$ from $\{\mathbf{Z}^{(t-1)}, \mathbf{Z}^{(t)}, \mathbf{Z}^{(t+1)}\}$ using the $v\varepsilon$ algorithm:

  $$\mathrm{vec}\dot{\mathbf{Z}}^{(t-1)} = \mathrm{vec}\mathbf{Z}^{(t)} + \left[\left[\Delta\mathrm{vec}\mathbf{Z}^{(t)}\right]^{-1} - \left[\Delta\mathrm{vec}\mathbf{Z}^{(t-1)}\right]^{-1}\right]^{-1},$$

where $\text{vec}\mathbf{Z}^* = (\mathbf{z}_1^{*\top} \ \mathbf{z}_2^{*\top} \ \cdots \ \mathbf{z}_r^{*\top})^\top$, and check the convergence by

$$\left\| \text{vec}\Delta\dot{\mathbf{Z}}^{(t-2)} \right\|^2 < \delta,$$

where δ is a desired accuracy.

The $v\varepsilon$ acceleration for PRINCALS can also accelerate the convergence of $\{\mathbf{Z}^{(t)}\}_{t \geq 0}$ preserving the properties of PRINCLAS. When the algorithm terminates, the estimate of $\mathbf{Z}$ is the final value of $\{\mathbf{Z}^{(t)}\}_{t \geq 0}$. From the Estimation of category quantifications step, we can obtain the estimate of $\mathbf{W}$ and then find the optimal scaled vector $\mathbf{y}^*$ for variable j in $\mathbf{J}_S$ using

$$\mathbf{y}_j^* = \mathbf{G}_j\mathbf{q}_j,$$

where $\mathbf{q}_j$ is obtained from the estimate of $\mathbf{W}_j$.

7.4 Numerical Experiments

In order to demonstrate the advantage of the $v\varepsilon$-ALS algorithm over the ordinal ALS algorithm, we examine their performances in terms of the number of iterations and CPU time (in seconds) required for convergence. All computations are performed with the statistical package R (R Development Core Team 2013) running on a Core i5 3.3 GHz computer with 4 GB of memory. The consumed CPU times are measured by the function proc.time.[1] Here PRINCIPALS is used as the ALS algorithm for nonlinear PCA. For all experiments, we set $\delta = 10^{-10}$ for convergence of the $v\varepsilon$-ALS algorithm and $|\theta^{*(t+1)} - \theta^{*(t)}| < 10^{-10}$ for the ALS algorithm, where $\theta^{*(t)}$ is the t-th update of θ^* calculated from

$$\theta^{*(t)} = \left\| \mathbf{Y}^{*(t)} - \mathbf{Z}^{(t)}\mathbf{A}^{(t)\top} \right\|^2.$$

7.4.1 Convergence of the Vε-ALS Algorithm Using Real Data

We use the data of Table 7.1 obtained from a teacher evaluation of 56 students and the following 13 questions (Q1–Q13) with 5 levels each; the lowest evaluation level is 1 and the highest 5.

---

[1] Times are typically available to 10 msec.

Table 7.1 Teacher evaluation data

Student	Q1	Q2	Q3	Q4	Q5	Q6	Q7	Q8	Q9	Q10	Q11	Q12	Q13
1	5	5	5	5	5	5	3	2	5	3	1	3	5
2	5	5	5	3	3	3	2	3	3	4	5	3	5
3	5	3	4	3	4	4	5	5	3	3	4	3	4
4	4	5	4	5	5	3	3	3	3	5	3	4	5
5	5	5	5	5	5	5	5	5	5	5	5	5	5
6	5	5	5	5	5	4	5	4	4	5	5	4	5
7	5	3	3	3	3	3	5	3	3	5	5	5	5
8	5	5	5	5	5	3	4	5	5	5	5	5	5
9	5	4	5	5	4	3	4	5	3	5	5	5	5
10	5	5	5	5	5	3	5	4	5	5	5	5	5
11	5	5	5	3	5	5	5	5	5	5	5	5	5
12	5	5	4	2	4	4	4	3	5	4	2	3	5
13	5	5	5	4	5	3	5	5	5	5	5	5	5
14	5	5	5	3	3	3	5	3	4	5	5	5	5
15	5	5	5	5	5	5	4	5	3	5	5	4	5
16	5	3	3	5	5	5	5	3	3	3	5	5	5
17	5	5	5	5	5	5	5	5	5	5	5	5	5
18	5	5	5	5	5	5	5	5	4	5	5	5	5
19	3	4	5	4	5	3	2	5	2	4	5	4	5
20	5	5	5	5	5	5	5	5	5	5	5	5	5
21	5	5	5	3	3	5	3	3	5	5	5	4	5
22	5	5	5	5	5	5	5	5	4	5	5	5	5
23	5	3	4	5	4	4	3	3	5	5	3	4	4
24	5	5	5	5	4	5	5	5	4	5	5	5	5
25	5	5	5	4	4	5	5	5	5	5	5	4	5
26	5	5	4	5	5	5	5	5	5	4	5	5	5
27	5	5	5	5	5	5	5	4	3	5	5	5	5
28	4	5	5	3	3	3	5	3	4	3	4	4	4
29	5	4	4	3	2	3	4	3	3	3	3	3	4
30	5	5	5	5	5	4	5	5	4	5	5	5	5
31	5	4	5	5	5	4	5	4	5	4	4	4	5
32	5	4	5	4	4	5	5	5	3	4	4	4	5
33	5	1	3	2	3	3	3	2	4	3	3	2	4
34	5	3	5	5	4	3	5	5	2	5	5	4	5
35	3	5	5	3	4	3	5	3	3	4	5	5	5
36	5	5	5	5	5	4	5	5	5	5	5	5	5
37	5	5	5	5	5	5	5	5	5	5	5	5	5

(continued)

Table 7.1 (continued)

Student	Q1	Q2	Q3	Q4	Q5	Q6	Q7	Q8	Q9	Q10	Q11	Q12	Q13
38	5	5	5	5	5	5	4	5	4	5	5	5	4
39	5	5	5	3	3	5	5	5	4	5	5	5	5
40	5	5	5	3	5	3	4	3	3	4	3	5	5
41	5	5	5	5	4	5	5	3	5	2	1	1	3
42	5	5	5	5	5	5	5	5	3	5	5	5	5
43	5	4	5	5	5	4	5	3	3	5	5	5	5
44	5	4	5	5	3	4	5	4	3	5	5	5	4
45	4	4	4	4	4	3	5	3	3	5	4	5	4
46	5	5	5	3	4	3	3	4	4	5	5	5	5
47	5	5	5	5	5	4	4	5	5	5	4	4	5
48	4	3	3	4	4	3	4	3	2	3	2	2	3
49	5	5	5	5	5	5	5	5	4	5	5	4	5
50	5	5	5	5	5	5	5	4	4	5	5	5	5
51	5	5	5	5	5	3	4	4	4	5	5	5	5
52	5	5	5	5	5	5	5	5	5	5	5	5	5
53	5	4	5	5	5	3	4	4	3	4	5	5	5
54	4	5	4	4	4	5	5	4	5	5	3	1	5
55	5	5	5	5	3	3	5	5	5	5	5	5	5
56	5	3	5	2	4	3	5	5	2	5	5	5	5

Q1: Loudness of the voice
Q2: Speed of the speaking
Q3: Clearness of the speaking
Q4: Size of the characters and figures on the blackboard
Q5: Carefulness and accuracy of the characters and figures on the blackboard
Q6: Accordance with the syllabus
Q7: Punctuality
Q8: Suitability of the textbook
Q9: Caution to whispers
Q10: Devised works to make contents easier to understand
Q11: Progress of the classwork
Q12: Consideration to students' abilities
Q13: Eagerness

In the experiments, we set $r = 3$ and study the convergence behaviors of the ALS and $v\varepsilon$-ALS algorithms. For the data, the ALS and $v\varepsilon$-ALS algorithms converge after 421 and 173 iterations, respectively. We examine the convergence behaviors of the $v\varepsilon$-ALS algorithm. Figure 7.1 illustrates the traces of $\{\log_{10} \|\mathbf{Y}^{*(t)} - \mathbf{Y}^{*(\infty)}\|^2\}_{1 \le t \le 173}$ for the ALS algorithm and $\{\log_{10} \|\dot{\mathbf{Y}}^{*(t)} - \mathbf{Y}^{*(\infty)}\|^2\}_{1 \le t \le 173}$ for the $v\varepsilon$-ALS algorithm till the convergence of the $v\varepsilon$-ALS algorithm attains. The figure shows that, when

Fig. 7.1 Convergence
behavior of the ALS and
$v\varepsilon$-ALS algorithms till the
convergence of the $v\varepsilon$-ALS
algorithm attains

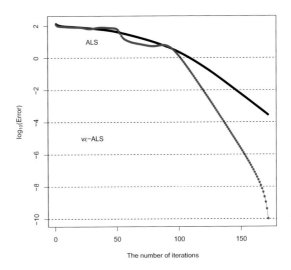

the sequence generated by the $v\varepsilon$-ALS algorithm converges after 173 iterations with $\delta = 10^{-10}$, the sequence of the ALS algorithm does not match the four-digit precision to $\mathbf{Y}^{*(\infty)}$. The figures demonstrate that $\{\dot{\mathbf{Y}}^{*(t)}\}_{t\geq 0}$ converges to $\mathbf{Y}^{*(\infty)}$ significantly faster than $\{\mathbf{Y}^{*(t)}\}_{t\geq 0}$.

Next we measure the rates of convergence of the ALS and $v\varepsilon$-ALS algorithms. The rates of convergence of these algorithms are assessed as

$$\tau = \lim_{t\to\infty} \tau^{(t)} = \lim_{t\to\infty} \frac{\|\mathbf{Y}^{*(t)} - \mathbf{Y}^{*(t-1)}\|}{\|\mathbf{Y}^{*(t-1)} - \mathbf{Y}^{*(t-2)}\|} \quad \text{for the ALS algorithm,}$$

$$\dot{\tau} = \lim_{t\to\infty} \dot{\tau}^{(t)} = \lim_{t\to\infty} \frac{\|\dot{\mathbf{Y}}^{*(t)} - \dot{\mathbf{Y}}^{*(t-1)}\|}{\|\dot{\mathbf{Y}}^{*(t-1)} - \dot{\mathbf{Y}}^{*(t-2)}\|} \quad \text{for the } v\varepsilon\text{-ALS algorithm.}$$

If the inequality $0 < \dot{\tau} < \tau < 1$ holds, we can say that $\{\dot{\mathbf{Y}}^{*(t)}\}_{t\geq 0}$ converges faster than $\{\mathbf{Y}^{*(t)}\}_{t\geq 0}$. Figure 7.2 indicates that $\{\dot{\mathbf{Y}}^{*(t)}\}_{t\geq 0}$ converges faster than $\{\mathbf{Y}^{*(t)}\}_{t\geq 0}$ in comparison with τ and $\dot{\tau}$. Next we investigate the speed of convergence of the $v\varepsilon$-ALS algorithm by

$$\dot{\rho} = \lim_{t\to\infty} \dot{\rho}^{(t)} = \lim_{t\to\infty} \frac{\|\dot{\mathbf{Y}}^{*(t)} - \mathbf{Y}^{*(\infty)}\|}{\|\mathbf{Y}^{*(t+2)} - \mathbf{Y}^{*(\infty)}\|} = 0. \tag{7.2}$$

From Brezinski and Zaglia (1991), if $\{\dot{\mathbf{Y}}^{*(t)}\}_{t\geq 0}$ converges to the same limit point $\mathbf{Y}^{*(\infty)}$ as $\{\mathbf{Y}^{*(t)}\}_{t\geq 0}$ and Eq. (7.2) holds, we say that $\{\dot{\mathbf{Y}}^{*(t)}\}_{t\geq 0}$ accelerates the convergence of $\{\dot{\mathbf{Y}}^{*(t)}\}_{t\geq 0}$. Figure 7.3 is the trace plot of $\{\dot{\rho}^{(t)}\}_{1\leq t\leq 173}$. The figure shows that the $v\varepsilon$-ALS algorithm accelerates the convergence of $\{\mathbf{Y}^{*(t)}\}_{t\geq 0}$. These figures demonstrate that the $v\varepsilon$-ALS algorithm significantly improves the rate of convergence of the ALS algorithm.

Fig. 7.2 Trace plots of
$\{\tau^{(t)}\}_{1 \leq t \leq 173}$ and
$\{\dot{\tau}^{(t)}\}_{1 \leq t \leq 173}$

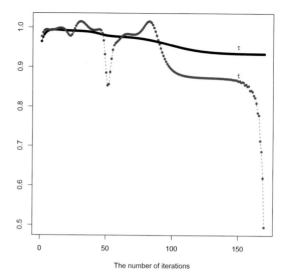

The number of iterations

Fig. 7.3 Trace plot of
$\{\dot{\rho}^{(t)}\}_{1 \leq t \leq 173}$

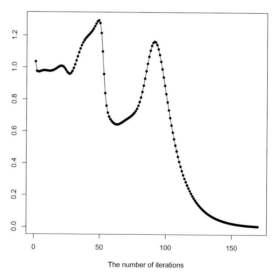

The number of iterations

7.4.2   Speed of Convergence of the Vε-ALS Algorithm Using Random Data

In this experiment, we study how much faster the vε-ALS algorithm converges than the ALS algorithm. We apply these algorithms to a random data matrix of 200 observations on 40 nominal variables with 10 levels and estimate **Z** and **A** on 5 (=r) principal components and **Y***. We replicate the computation 1000 times and measure the number of iterations and CPU time consumed.

Table 7.2 The number of iterations and CPU time of the ALS and vε-ALS algorithms

	The number of iterations		CPU time	
	ALS	vε-ALS	ALS	vε-ALS
Min.	106.0	40	3.12	1.480
1st Qu.	366.8	116	9.95	3.565
Median	514.0	164	13.82	4.895
Mean	676.0	219	18.09	6.406
3rd Qu.	751.2	244	20.01	7.090
Max.	10000.0	2754	262.19	76.370

Fig. 7.4 Boxplots of the numbers of iterations and CPU times of the ALS and vε-ALS algorithms

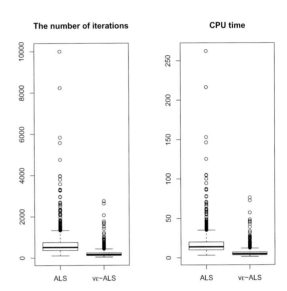

Table 7.2 shows the summaries of the number of iterations and CPU time taken by the ALS and vε-ALS algorithms. Figure 7.4 shows the boxplots of the numbers of iterations and CPU times of these algorithms. We see from the table and figure that the ALS algorithm requires more iterations and takes a longer time than the vε-ALS algorithm.

To compare the speed of convergence of the ALS and vε-ALS algorithms, we calculate iteration and CPU time speed-ups. The iteration speed-up is defined as

$$\frac{\text{The number of iterations of the ALS algorithm}}{\text{The number of iterations of the } v\varepsilon\text{-ALS algorithm}}.$$

The CPU time speed-up is calculated similarly to the iteration speed-up. Table 7.3 provides the speed-ups of the ALS and vε-ALS algorithms. We see from the first and second columns in the table that the vε-ALS algorithm converges about 3 times faster than the ALS algorithm in terms of the mean number of iterations and the

Table 7.3 Iteration and CPU time speed-ups of the ALS and $v\varepsilon$-ALS algorithms, and CPU time ratio

	Speed-up		CPU time ratio
	Iteration	CPU time	ALS/$v\varepsilon$-ALS
Min.	1.194	1.131	0.7867
1st Qu.	2.702	2.442	0.8809
Median	3.187	2.826	0.9007
Mean	3.223	2.890	0.8985
3rd Qu.	3.687	3.290	0.9202
Max.	8.027	7.233	0.9594

Fig. 7.5 Scatterplots of the iterations and CPU time speed-ups by the number of iterations of the ALS algorithm

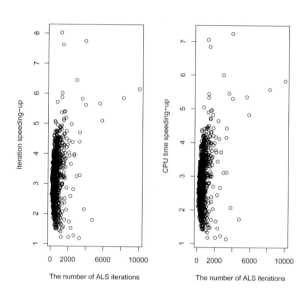

mean CPU time. Figure 7.5 shows the scatterplots of the iteration and CPU time speed-ups by the number of iterations of the ALS algorithm. The figure shows that the speed of convergence of the $v\varepsilon$-ALS algorithm is much faster than that of the ALS algorithm whereas the ALS algorithm requires many iterations for convergence. We can demonstrate this fact by drawing the boxplots of Fig. 7.6. Here, "less" ("more") means that the number of iterations of the ALS algorithm is less (more) than the 3rd quantile in Table 7.2. We see from the figure that, for the larger number of iterations of the ALS algorithm, the $v\varepsilon$ acceleration works well to greatly speed up the convergence of $\{\mathbf{Y}^{*(t)}\}_{t \geq 0}$. The result indicates that the $v\varepsilon$-ALS algorithm is more useful whereas the ALS algorithm takes the larger number of iterations.

The computation time per iteration of the $v\varepsilon$-ALS algorithm is longer than that of the ALS algorithm due to computation of the Acceleration step. We calculate the CPU time ratio of the ALS algorithm to the $v\varepsilon$-ALS algorithm. The ratio is defined by

Fig. 7.6 Boxplots of the iteration and CPU time speed-ups: "less" ("more") means that the number of iterations of the ALS algorithm is less (more) than the 3rd quantile in Table 7.2

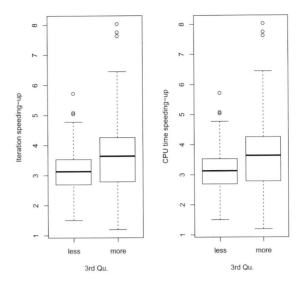

$$\left. \frac{\dfrac{\text{CPU time of the ALS algorithm}}{\text{the number of iterations of the ALS algorithm}}}{\dfrac{\text{CPU time of the } v\varepsilon\text{-ALS algorithm}}{\text{the number of iterations of the } v\varepsilon\text{-ALS algorithm}}} \right..$$

The third column of Table 7.3 is the summary of the CPU time ratio. The values in the table mean that the computation time of the ALS step accounts for 79–96% of that of the $v\varepsilon$-ALS algorithm. This means that the computational cost of the $v\varepsilon$ acceleration in the Acceleration step is less expensive than that of the ALS algorithm. Therefore the $v\varepsilon$ acceleration enables a significant reduction of the number of iterations and computation time with less computational effort.

We conclude from these results that the advantage of the $v\varepsilon$ acceleration is obvious.

7.4.3 Application of the Vε-ALS Algorithm to Variable Selection Problems

We consider variable selection problems for which the ALS algorithm requires a large amount of computation time and evaluate the performance of the $v\varepsilon$-ALS algorithm. In the variable selection for providing a simple interpretation of principal components, we find the best subset of variables consisting of q variables among p original variables ($1 \leq q \leq p$). Modified PCA (M.PCA) proposed by Tanaka and Mori (1997) derives principal components that are computed by a linear combination of a subset of variables but can reproduce all variables very well. When applying M.PCA to qualitative data, the ALS algorithm can be used for quantifying qualita-

Table 7.4 The numbers of iterations and CPU times of the ALS and vε-ALS algorithms and their speed-ups in application to variable selection for finding a subset of q variables

q	ALS		vε-ALS	
	Iteration	CPU time	Iteration	CPU time
Backward elimination				
20	2127	15.24	482	3.52
19	27,969	198.55	9499	71.08
18	17,066	121.34	6161	46.32
17	11,126	78.92	4768	35.86
16	12,481	87.91	4267	31.99
15	14,635	102.68	4421	32.96
14	11,585	81.13	3894	29.00
13	7930	55.52	2372	18.01
12	13,459	93.11	4442	32.62
11	8519	58.97	3415	25.18
10	8596	59.14	2433	18.16
9	12,077	82.53	3749	26.81
8	7409	50.75	2541	18.63
7	16,489	112.05	7026	50.08
6	28,333	191.91	19,695	138.63
5	36,108	243.76	11,840	83.54
4	402	3.02	131	1.35
Total	236,311	1636.53	91,136	663.74
Forward selection				
4	981,970	6985.23	270,478	2293.83
5	74,997	510.21	37,295	265.42
6	40,815	280.08	14,145	101.17
7	45,555	312.92	28,373	202.73
8	29,540	203.41	23,081	165.84
9	10,170	70.56	3058	22.70
10	9448	65.81	3574	26.43
11	7459	51.95	2775	20.65
12	5675	39.62	2124	15.99
13	4091	28.83	1403	10.75
14	6160	43.37	1648	12.42
15	7356	51.84	1603	12.15
16	8982	63.24	1951	14.55
17	2895	20.43	1042	7.84
18	2802	19.72	847	6.39
19	2016	14.21	541	4.10
20	707	4.99	229	1.74
Total	1,240,638	8766.42	394,167	3184.70

tive data, see Mori et al. (1997). Then Backward, Backward-forward, Forward and Forward-backward selections are utilized as cost-saving stepwise variable selection procedures. Chapter 4 presents these descriptions in detail.

We apply the ALS and vε-ALS algorithms to variable selection in M.PCA of qualitative data using simulated data that consist of 100 observations of 20 nominal variables with 10 levels.

Table 7.4 shows the number of iterations and CPU time taken by Backward and Forward selection procedures for finding a subset of q variables based on 4 ($=r$) principal components. The last row in the table shows the total number of iterations and total CPU time. The ALS algorithm finds the subsets after 236,311 iterations in Backward elimination and 1,240,638 iterations in Forward selection. The vε-ALS algorithm requires 91,136 and 394,167 iterations for these computations. The total computation times of the ALS algorithm for these variable selection procedures are 27.3 min and 2.4 h, respectively, while the vε-ALS algorithm takes only 11 and 53 min for the convergence; hence, computation time can be reduced to 41%($=663.74/1636.53$) and 36%($=3184.70/8766.42$) of those of the ALS algorithm. These results demonstrate that the vε acceleration step works well to accelerate the convergence of the sequence from the ALS step and consequently results in greatly decreased computation times in variable selection problems.

References

Brezinski, C., Zaglia, M.: Extrapolation Methods: Theory and Practice. Elsevier Science Ltd., North-holland, Amsterdam (1991)

Gifi, A.: Nonlinear Multivariate Analysis. Wiley, Chichester (1990)

Kuroda, M., Mori, Y., Iizuka, M., Sakakihara, M.: Accelerating the convergence of the EM algorithm using the vector epsilon algorithm. Comput. Stat. Data Anal. 55, 143–153 (2011)

Kruskal, J.B.: Nonmetric multidimensional scaling: A numerical method. Psychometrika, 29, 115–129 (1964)

Mori, Y., Tanaka, Y., Tarumi, T.: Principal component analysis based on a subset of variables for qualitative data. Data Science, Classification, and Related Methods (Proceedings of IFCS-96), pp. 547-554, Springer, Berlin (1997)

R Core Team: R: A language and environment for statistical computing. R Foundation for Statistical Computing, Vienna, Austria. ISBN 3-900051-07-0, http://www.R-project.org/ (2013)

Tanaka, Y., Mori, Y.: Principal component analysis based on a subset of variables: variable selection and sensitivity analysis. Am. J. Math. Manag. Sci. 17, 61–89 (1997)

Young, F.W., Takane, Y., de Leeuw, J.: Principal components of mixed measurement level multivariate data: an alternating least squares method with optimal scaling features. Psychometrika 43, 279–281 (1978)

Wynn, P.: Acceleration techniques for iterated vector and matrix problems. Math. Comput. 16, 301–322 (1962)

Index

© The Author(s) 2016
Y. Mori et al., *Nonlinear Principal Component Analysis and Its Applications*,
JSS Research Series in Statistics, DOI 10.1007/978-981-10-0159-8